Lecture Notes in Computer Science 559

Edited by G. Goos and J. Hartmanis

G. Butler

Fundamental Algorithms for Permutation Groups

Springer-Verlag
Berlin Heidelberg New York
London Paris Tokyo
Hong Kong Barcelona
Budapest

Author

Gregory Butler
CICMA
Department of Computer Science, Concordia University
1455 de Maisonneuve Blvd. West
Montreal, Quebec H3G 1M8, Canada

CR Subject Classification (1991): I.1.2, F.2.2, G.2.1-2

ISBN 3-540-54955-2 Springer-Verlag Berlin Heidelberg New York
ISBN 0-387-54955-2 Springer-Verlag New York Berlin Heidelberg

Typesetting: Camera ready by author
Printing and binding: Druckhaus Beltz, Hemsbach/Bergstr.
45/3140-543210 - Printed on acid-free paper

Preface

These notes are derived from an 18-hour lecture course on Symbolic and Algebraic Computation that I have been giving to Honours computer science students at the University of Sydney since 1983, and lectures to mathematics and computer science students at the University of Bayreuth in 1990. Due to the very wide scope of the field, I concentrate on my area of speciality — algorithms for permutation groups. The (short) course at Sydney emphasises Chapters 2, 3, 4, 7, 10, and 11.

The aim of the course is to develop each of the algorithms from scratch, showing how each new piece of group-theoretical information can be used to improve the algorithm. As such, it could be regarded as a course in algorithm development. I assume no background in group theory, so the piecewise introduction of information allows students to become familiar with one concept at a time. Another advantage of this approach is that the correctness of the algorithm is justified as it is developed.

The examples and the exercises assist students to learn the group theory and the working of the algorithms. Occasionally, they show how the algorithms could be further improved, or they develop an alternate algorithm. Access to the computer algebra system Cayley would be beneficial. Students could then easily study other examples and implement the algorithms.

The bibliographical remarks explain the history of the algorithms, and often place them in the broader context.

The algorithms are presented using Pascal control structures with some exceptions. We use the **for**-loop of Pascal to run over sets, even sets that increase during the loop's execution. The meaning of such a loop is that each set member is to be considered precisely once, including those members added during the loop's execution. The order in which they are considered is generally unimportant, but for precision, assume it is the same as the order in which they were added to the set. We do not use **begin** ... **end** to form compound statements. Instead all the control statements have a corresponding closing bracket such as **end if**, **end for**, and **end while** to indicate the extent of the statement. In addition, we occasionally use an **exit**-statement to terminate an algorithm, a **return**-statement to terminate a procedure, and a **result**-statement to terminate a function. The algorithms also use English and mathematical statements and operators. (* Comments in the algorithms are enclosed like this sentence. *)

Some general assumptions are made during the analysis of algorithms. In Chapters 3-6, the analysis of the algorithms is in terms of element multiplications and searches of lists of elements. To this point, the analyses are independent of the element representation. However, we sometimes obtain a single total cost based on multiplications that assumes the elements are permutations. In this case a comparison of elements costs the same as a multiplication, and we assume a (hash) search requires at most two comparisons. In Chapters 7-14, the analyses are concerned with the accesses made to permutations, sets, and Schreier vectors. In essence, we assume that the simple variables of an algorithm will be stored in registers or very fast memory, and therefore the time to access them will be negligible. The cost of accessing one entry, or storing the value in one entry, of a permutation, set, or Schreier vector will be counted as one operation. We are viewing these as one read or write to memory. In any case, these accesses and stores will constitute the bulk of the cost of an algorithm.

The first part of the book, comprising Chapters 2-6, is an introduction to computational group theory for those without a knowledge of group theory. Even so, some mathematical sophistication is a distinct advantage. For example, familiarity with geometry, manipulation of algebraic formulae, and straightforward analysis of algorithms is desirable. Those with a knowledge of group theory, or even a knowledge of computational group theory will find something of interest. We have attempted to develop the algorithms from first principles showing where the particular pieces of group theoretical knowledge have influenced the algorithm, and how they can be improved by using this knowledge, until we reach the state of the art. We have attempted to justify (prove?) the algorithms as we develop them, and to analyse their time and space usage. For many of the algorithms a complete analysis is not known, so we have fallen short on our last goal.

This part is really a slow introduction to group theory and uses of the elementary concepts of group theory in algorithms that handle small groups — that is, groups for which we can store a list of all their elements. It serves also as a first exposure to some techniques that are relevant to large permutation groups, and of course, as a source of information about algorithms for small groups. The problems tackled in this part are: determining a list of elements from a generating set; searching a group for elements with a given property; determining defining relations from a Cayley graph; and determining the lattice of all subgroups.

The first two problems cover the essential fundamental algorithms. The third problem introduces the Cayley graph, which is a useful representation of a group, and provides a link with combinatorial group theory and its algorithms. The last problem, besides being historically significant, leads to a beautifully subtle algorithm that effectively tackles what at first sight appears an impossibly large task.

All our examples are groups of permutations. The analysis of the algorithms is in terms of element multiplications and searches of lists of elements. To this point, the analyses are independent of the element representation. However, we sometimes obtain a single total cost based on multiplications that assumes the elements are permutations. In this case a comparison of elements costs the same as a multiplication, and we assume a (hash) search requires at most two comparisons.

The second part of the book comprises Chapters 7-14 and considers algorithms specific to permutation groups. One benefit of restricting to permutation groups is that the algorithms can handle very large groups, say of order 10^{20}, and of degree up to 10 000. For some problems there are specialist techniques to handle groups of degree up to 10^5, but we will not discuss those here.

The power of these algorithms comes from intimate use of how the permutations act on the points. Think of it as an attempt to make the complexity of the algorithms a function of the degree rather than a function of the group order. The action on points can often be conveniently represented by orbits and Schreier vectors. We will discuss these first, and then present two elementary uses of the information contained in the orbits and Schreier vectors. The two uses are determining whether a group is regular (that is, whether any element of the group fixes a point), and whether a group is imprimitive (that is, whether the group leaves invariant a (non-trivial) partition of the points). In the end, both these uses simply use the information provided by the generators of the group.

A chain of subgroups, where each subgroup fixes at least one more point than the previous one, provides the inductive basis for the remaining algorithms. Such a chain is called a stabiliser chain. Associated with a stabiliser chain is a base and strong generating set — *strong* because it contains generators for each subgroup in the chain, and not just for the whole group. We can represent all the elements of the group, and effectively compute with them, by a base, a strong generating set, and a Schreier vector for each subgroup in the stabiliser chain.

After introducing the concepts of stabiliser chain, base, and strong generating set, we will discuss at length how elements can be represented, and how we could, if necessary, generate all the elements of the group. We delay discussing how one determines a base and strong generating set for a permutation group until the reader has had more experience with the concepts.

A major aim of this part is to discuss searching very large permutation groups. A backtrack algorithm is used. The searching of permutation groups provides a good forum for discussing the various techniques for improving the heuristics of a backtrack algorithm. Several improvements are applicable to all backtrack searches of permutation groups, while others, even though the details rely on what we are searching for, demonstrate the universality of choosing a base appropriate to the problem, and of preprocessing information that is repeatedly required during the search. The improvements applicable to all backtrack searches of permutation groups are just another form of discarding the elements in a coset. We have met this strategy already when searching small groups.

After choosing an appropriate base for the search, we require a corresponding strong generating set. The base change algorithm will provide one from an existing base and strong generating set. This algorithm is presented.

Now we can painlessly present an algorithm that determines a base and strong generating set of a permutation group from a set of generators. We present the Schreier-Sims algorithm. There are many variations of this algorithm. They are generally called Schreier methods or Schreier-Sims methods. We will mention a few variations in passing.

The analyses of this part make some general assumptions. They are concerned with the accesses made to permutations, sets, and Schreier vectors. In essence, we assume that the simple variables of an algorithm will be stored in registers or very fast memory, and therefore the time to access them will be negligible. The cost of accessing one entry, or storing the value in one entry, of a permutation, set, or Schreier vector will be counted as one operation. We are viewing these as one read or write to memory. In any case, these accesses and stores will constitute the bulk of the cost of an algorithm.

The third part of the book comprising Chapters 15-18 looks at the role of homomorphisms in algorithms for permutation groups. In particular, Chapter 16 discusses the computation of Sylow subgroups by using homomorphisms; and Chapter 18 discusses the conversion from a permutation representation to a power-commutator presentation when the group is a p-group or a soluble group, so that these special cases can utilise the efficient algorithms for soluble groups and p-groups. Chapter 17 introduces the notion of a power-commutator presentation and some elementary algorithms based on that representation.

At this stage, the reader will have met the fundamental algorithms for handling permutation groups.

The last chapter briefly discusses what has been omitted from this book and gives pointers to the relevant literature.

I would like to thank all my students. They were the guinea pigs as I experimented with this approach, often pointing out errors and the need for further clarification. In particular, Jowmee Foo and Peter Merel carefully read the first draft, and Bernd Schmalz made detailed notes of the lectures at Bayreuth. Volkmar Felsch helped enormously in clarifying some historical aspects. John Cannon, Volkmar Felsch, Mike Newman, Jim Richardson, and Charles Sims have made numerous comments that have greatly improved the text. They have been conscientious, thorough, and detailed. I thank them all, and accept responsibility for any remaining errors or shortcomings.

Montreal
September 1991

Greg Butler

Contents

Part 1: Small Groups

Part 2: Permutation Groups

Chapter 1. Introduction

Computational group theory is primarily concerned with algorithms which investigate the structure of groups. This includes the development of algorithms, their proof of correctness, an analysis of their complexity, implementation of the algorithms, and determining the limits of their application in practice. The field is also concerned with the design of software systems which incorporate the algorithms, and allow their application to problems. This involves the design of a user language, the implementation of an interpreter or compiler, and the provision of decision procedures that choose appropriate default algorithms. These systems and algorithms assist in mathematical research in a wide range of disciplines including theoretical physics, topology, combinatorics, and group theory. The systems are also examples of computer-aided instruction, and embody a wealth of knowledge in the algorithms, and in various libraries of examples.

The field poses many challenging problems. Some of the tasks are provably undecidable. Others are known to be hard in the complexity sense. For many, their status is simply unknown, and any algorithm for the task is a breakthrough. The software problems are also very challenging. Each algorithm may require six months work to implement effectively, to study its performance, and to tune it. The engineering problems of large software systems have to be overcome if the system is to be robust, portable, maintainable, and rapidly modified in the light of new breakthroughs in algorithms and data structures. Yet despite these problems, computational group theory has been a very successful branch of computer algebra, and knowledge based systems.

Over the last decade, and certainly since the last major survey of Cannon(1969), there has been enormous development in the field of computational group theory. Here we bring the reader up to date with the major representations of group-theoretical information and some principles behind the design of algorithms which manipulate them. Systems for computational group theory are briefly surveyed, and some recent applications are presented.

The best general introductions to the field are the proceedings of the 1982 symposium at Durham edited by Atkinson(1984), and the survey of coset table algorithms by Neubüser(1982). An extensive and concise list of known algorithms is provided by Neubüser's article in Buchberger, Collins, and Loos(1982), while Felsch(1978), in its current version, is a complete bibliography of the field.

Machine Representation of Groups

Computation with groups demands that the elements of the groups, and the groups as a whole be represented in the computer by some data structure. Concrete representations of elements, for example, by permutations and matrices are preferred to abstract representations such as symbolic products of generators. The abstract representations often require a deductive approach to infer properties of elements, which is open to problems of undecidability and inefficiency. The critical operations for any representation of elements are multiplication of elements, and

determination of equality of two elements.

The representation of the group has a whole must address the following concerns.

Is the representation of the group compact (that is, space efficient)?

Is it easy (or at least possible) to compute the representation?

Is there a compact representation of an individual element?

Is membership in the group and/or subgroups easily determined?

Is it easy to determine the order of the group, and to enumerate its elements?

How easy is it to construct subgroups, quotients, images and preimages of homomorphisms?

The description of the group in the first place must be finite, so computational group theory is only concerned with *finitely generated* groups.

The most common descriptions of groups use

permutations,

matrices,

power-commutator-presentations, and

finite presentations.

The first three provide concrete representations, while the last is abstract. Other machine representations are lists of elements, coset tables, and character tables. A coset table (when it can be computed) provides a concrete representation for groups given by finite presentations.

The effectiveness of algorithms varies from one representation to another. Hence, it is useful to be able to convert between representations whenever possible. This allows an appropriate representation to be chosen for each problem at hand.

Many of the algorithms for group theory rely on the *divide-and-conquer* paradigm. The problem for a group G is reduced to a problem about a subgroup H, or a quotient Q of G. The reductions can be applied recursively (or iteratively) when a *chain* of subgroups

$$G=G(1) \geq G(2) \geq \cdots \geq G(m+1)=\{identity\}$$

or normal subgroups, respectively, are known. Homomorphisms are another way in which quotients can be obtained for the reduction.

A major benefit of the three concrete representations of groups is that there is naturally a chain of subgroups (in the case of permutations or matrices) or normal subgroups (in the case of power-commutator-presentations).

A study of the divide-and-conquer paradigm in group theory by Butler(1986) also demonstrates the important role of algorithms for

subgroup construction,

coset enumeration, and

homomorphisms,

as building blocks for algorithms employing the paradigm.

Systems

The Cayley system, developed at the University of Sydney by John Cannon(1984), operates at over 150 sites in 21 countries. It is *the* system for computational group theory. The user language has Pascal control structures; facilities for defining algebraic objects; sets, sequences, and mappings of algebraic objects; and over 300 in-built algebraic functions. It handles those algebraic objects needed to define groups, or on which groups act, such as rings of integers, finite fields, vector spaces, modules, matrix rings, groups, polynomial rings, and incidence structures. Cayley allows all the representations of the previous sections, and has an extensive collection of the known algorithms. The system was originally developed in Fortran, but was ported to C in 1987. There are approximately 300,000 lines of code. The system has required about 45 man-years effort to date.

The system CAS was developed in Aachen by Joachim Neubüser and his colleagues(1984) for computations involving group characters. The system supports individual characters, partially defined character tables, complete character tables, arbitrary precision integers, and irrationalities. There are many built-in methods for generating complete character tables (including generic formulae of some Chevalley groups, and a library for the sporadic simple groups), generating individual characters, testing orthogonality relations, computing structure constants, and handling modular characters. The system was originally developed in Fortran, but has now been ported to C.

The SOGOS system was also developed at Aachen, see Laue et al(1984). This system supports soluble groups and p-groups represented by power-commutator-presentations. Many of the first implementations of algorithms manipulating pcp's were part of SOGOS. The system is implemented in Fortran and has a user interface similar to Cayley's.

There is also a Pascal system, CAMAC2 (Leon, 1984b), under development for the investigation of permutation groups and combinatorial objects.

Some Applications

Within mathematics, groups are ubiquitous. Applications arise in geometry, topology, combinatorial theory, number theory (especially algebraic number theory), and, of course, group theory. Space permits only a few examples. Several large simple groups were constructed by computer, most notably, the "Baby Monster" as a permutation group of degree 13 571 955 000 and order $2^{41}\ 3^{13}\ 5^{6}\ 7^{2}\ 11\ 13\ 17\ 19\ 23\ 31\ 47$ by Leon and Sims(1977), and Janko's group of order $2^{21}\ 3^{5}\ 5\ 7\ 11^{3}\ 23\ 31\ 37\ 43$ as a 112-dimensional matrix group over $GF(2)$ by Norton(1980) and others at Cambridge. Knots have been classified with the aid of computer (see Havas and Kovács,1984), as have certain 3-manifolds (by Richardson and

Rubenstein), *p*-groups (Ascione, Havas, and Leedham-Green, 1977), and crystallographic groups (Brown, Bülow, Neubüser, Wondratschek, and Zassenhaus, 1978) of interest in X-ray diffraction.

The Cayley system is widely used as a teaching aide. Its role at Sydney is described by Cannon and Richardson(1984/5).

The techniques for permutation groups are used in the programs of B.D. McKay(1981) and W. Kocay(1984), which are the cutting edge of practical general graph isomorphism programs. Leon(1979, 1982, 1984a) applies these techniques to determine automorphism groups of error correcting codes and Hadamard matrices, and also in their enumeration. Groups are often used in the construction and investigation of combinatorial objects and their extensions, for example, Magliveras and Leavitt(1984), Leon(1984a).

The period 1979-1981 saw major advances in the theoretical complexity of the graph isomorphism problem by Babai(1979), Luks(1980), and Hoffman(1982) which were based on group theoretic algorithms. Graph isomorphism is polynomial for graphs of bounded valence. Their work made the complexity of group theoretic problems fashionable, see Furst, Hopcroft, and Luks(1979), Even and Goldreich(1981), Babai, Kantor, and Luks(1983), Babai and Szemeredi(1984).

Magliveras, Oberg, and Surkan(1987) use "logarithmic signatures" of permutation groups - essentially a base and strong generating set - as a means of encrypting data, and as a highly effective random number generator.

Bibliographical Remarks

ASCIONE, J.A., HAVAS, G., and LEEDHAM-GREEN, C.R. (1977): A computer-aided classification of certain groups of prime power order, *Bull. Aust. Math. Soc.*, **17**, pp. 257-274, 317-320, microfiche supplement.

ATKINSON, M.D. (ed.) (1984): *Computational Group Theory* (Proc. LMS Symp. on Computational Group Theory, Durham, July 30-August 9, 1982), Academic Press, London.

BABAI, L. (1979): Monte Carlo algorithms in graph isomorphism testing, manuscript.

BABAI, L., KANTOR, W., and LUKS, E. (1983): Computational complexity and the classification of finite simple groups, *Proc. 24th IEEE Foundations of Computer Science*, pp.162-171.

BABAI, L., and SZEMEREDI, R. (1984): On the complexity of matrix group problems I, *Proc. 25th IEEE Foundations of Computer Science*, pp.229-240.

BROWN, H., BÜLOW, R., NEUBÜSER, J., WONDRATSCHEK, H., and ZASSENHAUS, H. (1978): *Crystallographic Groups of Four-Dimensional Space*, Wiley, New York.

BUCHBERGER, B., COLLINS, G.E., and LOOS, R. (1982): *Computer Algebra : Symbolic and Algebraic Computation*, Springer-Verlag, Wien.

BUTLER, G. (1986): Divide-and-conquer in computational group theory, **SYMSAC '86** (Proceedings of the 1986 ACM Symposium on Symbolic and Algebraic Computation, Waterloo, July 21-23, 1986), ACM, New York, pp.59-64.

CANNON, J.J. (1969): Computers in group theory: A survey, *Commun. ACM,* **12,** pp.3-12.

CANNON, J.J. (1984): An introduction to the group theory language, Cayley, in Atkinson(1984), pp.145-183.

CANNON, J.J., and RICHARDSON, J.S. (1984/5): Cayley - Teaching group theory by computer, *SIGSAM Bull.,* **18/19,** pp.15-18.

EVEN, S., and GOLDREICH, O. (1981): The minimum-length generator sequence problem is NP-hard, *J. Algorithms,* **2,** pp.311-313.

FELSCH, V. (1978): A bibliography on the use of computers in group theory and related topics: algorithms, implementations, and applications, *SIGSAM Bull.,* **12,** pp.23-86.

FURST, M., HOPCROFT, J.E., and LUKS, E. (1980): Polynomial-time algorithms for permutation groups, *Proc. 21st IEEE Foundations of Computer Science,* pp.36-41.

GORENSTEIN, D. (1985): *The enormous theorem,* Scientific American **253,** 6, pp.92-103.

HALL, Jr, M. (1959): *The Theory of Groups,* Macmillan, New York.

HAVAS, G., and KOVACS, L.G. (1984): Distinguishing eleven crossing knots, in Atkinson(1984), pp.367-373.

HOFFMAN, C.M. (1982): *Group Theoretic Algorithms and Graph Isomorphism,* Lecture Notes in Computer Science, **136,** Springer-Verlag, Berlin.

KOCAY, W.L. (1984): Abstract data types and graph isomorphism, *J. Combinatorics, Information, and Systems Science,* **9,** pp.247-259.

LAUE, R., NEUBÜSER, J., and SCHOENWAELDER, U. (1984): Algorithms for finite soluble groups and the SOGOS system, in Atkinson(1984), pp.105-135.

LEON, J.S. (1979): An algorithm for computing the automorphism group of a Hadamard matrix, *J. Comb. Theory (A),* **27,** pp.289-306.

LEON, J.S. (1982): Computing automorphism groups of error correcting codes, *IEEE Trans. Infor. Theory,* **IT-28,** pp.496-511.

LEON, J.S. (1984a): Computing automorphism groups of combinatorial objects, in Atkinson(1984), pp.321-335.

LEON, J.S. (1984b): CAMAC2: A portable system for combinatorial and algebraic computation, *EUROSAM '84* (Proc. Internat. Symp. Symbolic and Algebraic Computation, Cambridge, July 9-11, 1984), J. Fitch (ed.), *Lecture Notes in Computer Science,* **174,** Springer-Verlag, Berlin, pp.213-224.

LEON, J.S., and SIMS, C.C. (1977): The existence and uniqueness of a simple group generated by (3,4)-transpositions, *Bull. Amer. Math. Soc.,* **83,** pp.1039-1040.

LUKS, E. (1980): Isomorphism of graphs of bounded valence can be tested in polynomial time, *Proc. 21st IEEE Foundations of Computer Science,* pp.42-49.

MAGLIVERAS, S.S., and LEAVITT, D.W. (1984): Simple 6-(33,8,36) designs from $P\Gamma L_2(32)$, in Atkinson(1984), pp.337-352.

MAGLIVERAS, S.S., OBERG, B.A., and SURKAN, A.J. (1987): A new random number generator from permutation groups, *Rend. Sem. Mat. Fis. Milano,* **54,** pp. 203-223.

MAGNUS, W., KARRASS, A., and SOLITAR, D. (1966): *Combinatorial Group Theory: Presentations of Groups in Terms of Generators and Relations,* Wiley Interscience, New York.

MCKAY, B.D. (1981): Practical graph isomorphism, (Proc. Tenth Manitoba Conf. on Numerical Math. and Computing, Winnipeg, vol. 1), *Congr. Numer.,* **30,** pp. 45-87.

NEUBÜSER, J. (1982): An elementary introduction to coset table methods in computational group theory, *Groups - St Andrews 1981* (Proc. internat. conf., St Andrews, July 25-August 8, 1981), London Mathematical Society Lecture Note Series, **71**, C.M. Campbell and E.F. Robertson (eds), Cambridge University Press, Cambridge, pp.1-45.

NEUBÜSER, J., PAHLINGS, H., and PLESKEN, W. (1984): CAS; Design and use of a system for the handling of characters of finite groups, in Atkinson(1984), pp.195-247.

NORTON, S.P. (1980): The construction of J_4, *Proc. Symp. Pure Math.,* **37,** pp.271-277.

WIELANDT, H. (1964): *Finite Permutation Groups,* Academic Press, New York.

Chapter 2. Group Theory Background

This chapter presents the background necessary for understanding the algorithms for small groups. This is very elementary group theory but our emphasis is slightly different from most group theory texts. The examples we discuss will be permutation groups, because the reader will need to be familiar with permutation groups for later chapters, and because we see no advantage in introducing other descriptions of groups at this stage. The concepts to be discussed are group, generators, permutation, subgroup, and coset.

Groups

A *group* is a non-empty set G (of elements) with a multiplication operator $\times$ with the properties that

(1) G is closed under $\times$ (that is, for all $g,\ h \in G, g \times h \in G$),
(2) $\times$ is associative (that is, for all $g,\ h,\ k \in G$, $(g \times h) \times k = g \times (h \times k)$),
(3) G has an identity id (that is, for all $g \in G$, $g \times id = id \times g = g$),and
(4) every element g of G has an inverse g^{-1} such that $g \times g^{-1} = g^{-1} \times g = id$.

The *order* of the group G is the number of elements in the set G. It is denoted by $|G|$. We will only be concerned with *finite* groups. In this case, for each element g there is always a smallest positive integer m such that

$$g^m = \underbrace{g \times g \times \cdots \times g}_{m \text{ times}} = id$$

The integer m is called the *order* of the element g. It is denoted by $|g|$. Note that $g^{-1} = g^{|g|-1}$.

Our first example is the symmetries of a square in Figure 1a. The rotations and reflections of a square form a group with eight elements. The multiplication operation composes the transformations from left to right. We list the elements in Figure 1b.

Figure 1a: A square

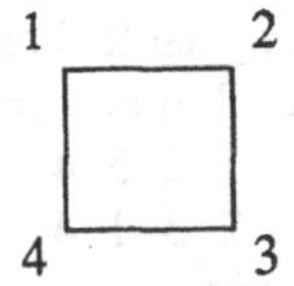

Note that $elt[1]$ is id, and that the order of $elt[2]$ is four. The first four elements are the rotations about the centre, while the fifth to eighth elements are the reflections of the first four about the leading diagonal.

Let S be a subset of a finite group G. The set S *generates* G if every element of G can be written as a product

Figure 1b : Symmetries of the Square

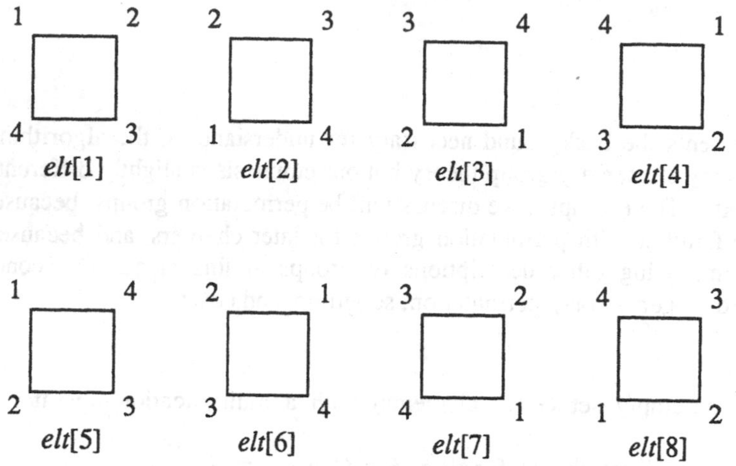

$$g = s_{i_1} \times s_{i_2} \times \cdots \times s_{i_m}$$

of elements in S, for some m dependent on g. We call S a *set of generators* for G and denote this by $G = \langle S \rangle$.

Let G be the symmetries of the square. Then G is generated by $\{elt\,[2], elt\,[5]\}$ since

$elt\,[1] = id$, is the empty product,
$elt\,[2] = elt\,[2]$,
$elt\,[3] = elt\,[2] \times elt\,[2]$,
$elt\,[4] = elt\,[2] \times elt\,[2] \times elt\,[2]$,
$elt\,[5] = elt\,[5]$,
$elt\,[6] = elt\,[2] \times elt\,[5]$,
$elt\,[7] = elt\,[2] \times elt\,[2] \times elt\,[5]$, and
$elt\,[8] = elt\,[2] \times elt\,[2] \times elt\,[2] \times elt\,[5]$.

We could also write $G = \langle elt\,[2], elt\,[5] \rangle$.

Permutations

A permutation is a bijection from a set to itself. For example, we can specify a bijection of the set {1,2,3,4} by listing the image of each member of the set, viz

1 2 3 4
2 3 1 4

A shorter notation is to omit the top line, viz

/2 3 1 4/

This is called the *image form* of the permutation.

The elements of the group of symmetries of the square as permutations of {1,2,3,4} are now listed in image form.

/1 2 3 4/	/4 1 2 3/	/3 4 1 2/	/2 3 4 1/
elt[1]	*elt*[2]	*elt*[3]	*elt*[4]
/1 4 3 2/	/2 1 4 3/	/3 2 1 4/	/4 3 2 1/
elt[5]	*elt*[6]	*elt*[7]	*elt*[8]

A *cycle* of a permutation is a sequence of set members

$$i_1, i_2, \cdots i_c$$

where

i_2 is the image of i_1,
i_3 is the image of i_2,
.
.
.
i_c is the image of i_{c-1}, and
i_1 is the image of i_c.

For example, 1, 2, 3 is a cycle of the permutation

/ 2 3 1 4 /.

A permutation may be specified by listing its cycles, viz

(1, 2, 3)(4)

Usually the cycles of length one are omitted, viz

(1, 2, 3)

This is called the *cycle form* of the permutation.

Multiplication of permutations is from left to right. First take the image under the left permutation, and then take the image of this under the right permutation. For example, if g = / 2 3 1 4 / and h = / 2 1 4 3 / then $g \times h$ = / 1 4 2 3 /, as shown in Figure 2.

Figure 2 : Multiplying Permutations

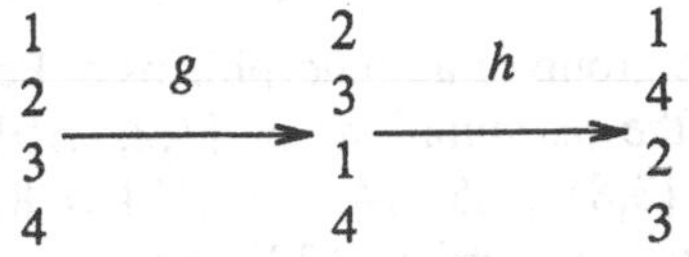

Our second example is the set of all permutations of {1,2,3,4}. This set is a group called the symmetric group of degree 4, (because the set {1,2,3,4} has size 4). There are 24 such permutations, so the order of the group is 24.

Our third example is the set of symmetries of the projective plane of order two. Consider a three-dimensional vector space over GF(2), the field of two elements {0,1}. The transformations which preserve the structure of a vector space are the linear transformations which map linearly independent vectors to linearly independent vectors. These are usually represented by *matrices*, in this case 3×3 matrices with entries from GF(2). There are 168 such invertible matrices, which form a group generated by

$$A = \begin{bmatrix} 0 & 1 & 0 \\ 1 & 0 & 0 \\ 0 & 0 & 1 \end{bmatrix} \quad B = \begin{bmatrix} 1 & 0 & 0 \\ 0 & 0 & 1 \\ 1 & 1 & 0 \end{bmatrix}$$

There is also a well-known geometry (Figure 3) associated with this vector space. The *projective plane* of order two is a set of seven points, together with seven lines, such that each line intersects in precisely one point, and each pair of points lie on a unique line. We can take the points to be the nonzero vectors, and the lines to be the two-dimensional subspaces. The 168 linear transformations preserve the geometry, in the sense that they map lines to lines. Regarded as permutations of the seven points, the group generators are

$$a = (1,2)\,(3,5) \qquad b = (2,4,3)\,(5,7,6)$$

Figure 3 : Projective Plane of Order Two

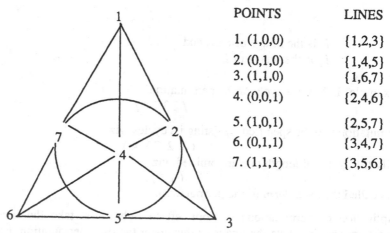

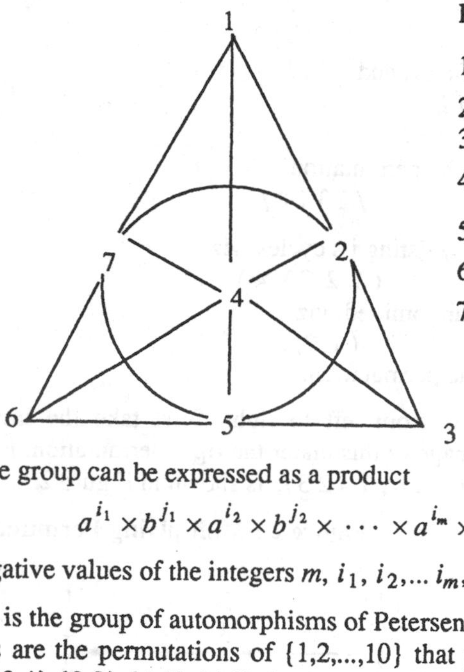

POINTS	LINES
1. (1,0,0)	{1,2,3}
2. (0,1,0)	{1,4,5}
3. (1,1,0)	{1,6,7}
4. (0,0,1)	{2,4,6}
5. (1,0,1)	{2,5,7}
6. (0,1,1)	{3,4,7}
7. (1,1,1)	{3,5,6}

Every element in the group can be expressed as a product

$$a^{i_1} \times b^{j_1} \times a^{i_2} \times b^{j_2} \times \cdots \times a^{i_m} \times b^{j_m}$$

for suitable non-negative values of the integers $m, i_1, i_2, \ldots i_m, j_1, j_2, \cdots j_m$.

Our fourth example is the group of automorphisms of Petersen's graph, shown in Figure 4. The automorphisms are the permutations of {1,2,...,10} that permute the edges {1,2} {1,5} {1,6} {2,3} {2,7} {3,4} {3,8} {4,5} {4,9} {5,10} {6,8} {6,9} {7,9} {7,10} {8,10} amongst themselves. The group has order 120 and is generated by / 1 2 3 8 6 5 7 4 10 9 /, / 1 5 4 3 2 6 10 9 8 7 /, and / 3 4 5 1 2 8 9 10 6 7 /.

Our fifth example is the set of all operations of Rubik's cube. These may be viewed as permutations of the starting configuration where we number the squares which are not in the centre of a face. The quarter turns of each face generate the group, so, with suitable numbering of the 48 noncentral squares, the generators are

Figure 4 : Petersen's Graph

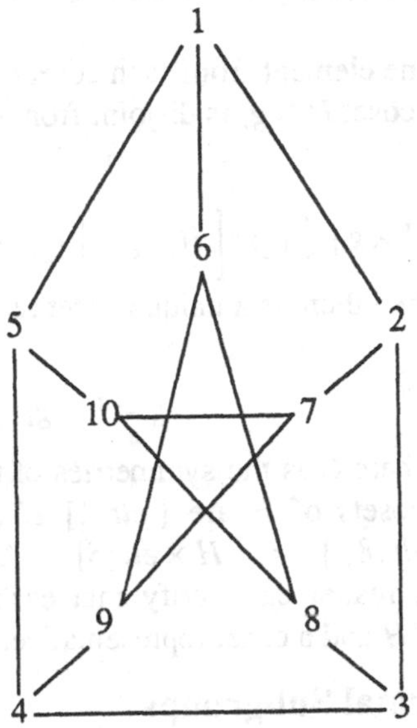

(1,3,8,6)(2,5,7,4)(9,48,15,12)(10,47,16,13)(11,46,17,14);
(6,15,35,26)(7,22,34,19)(8,30,33,11)(12,14,29,27)(13,21,28,20);
(1,12,33,41)(4,20,36,44)(6,27,38,46)(9,11,26,24)(10,19,25,18);
(1,24,40,17)(2,18,39,23)(3,9,38,32)(41,43,48,46)(42,45,47,44);
(3,43,35,14)(5,45,37,21)(8,48,40,29)(15,17,32,30)(16,23,31,22); and
(24,27,30,43)(25,28,31,42)(26,29,32,41)(33,35,40,38)(34,37,39,36).

The group has order

$$2^{27}3^{14}5^{3}7^{2}11$$
$$= 43\ 252\ 003\ 274\ 489\ 856\ 000.$$

This group is *not* small. It can, however, be handled by the algorithms for large permutation groups.

Subgroups and Cosets

A *subgroup* of a group G is a subset of G that is itself a group with the same multiplication operator as G. For example, the rotations of the square are a subgroup of order four of the symmetries of the square, and the symmetries of the projective plane of order two that leave the point 1 fixed form a subgroup of order 24 of the group of all symmetries of the plane.

Given a subgroup H of a group G and an element g of G, the *(right) coset* $H \times g$ is the set of elements $\{ h \times g \mid h \in H \}$. We can easily prove that if $\overline{g} \in H \times g$ then $H \times \overline{g} = H \times g$. This means that two cosets of H in G are either disjoint or the same. Thus the cosets of H partition the elements of G. Furthermore, all the cosets of H have size $|H|$. The number of cosets is called the *index* of H in G and is denoted by $|G:H|$. This gives

Lagrange's Theorem

If H is a subgroup of a finite group G then $|G| = |H| \times |G:H|$.

If we choose precisely one element from each coset of H, we get *a set of coset representatives* $\{g_1, g_2, \ldots, g_m\}$. Each coset $H \times g_i$ is disjoint from the others, so the elements of G (without repetition) are

$$\left[H \times g_1\right] \cup \left[H \times g_2\right] \cup \cdots \cup \left[H \times g_m\right]$$

If g is an element of G then there is a unique coset representative g_i and a unique element h of H such that

$$g = h \times g_i$$

Consider the example where G is the symmetries of the square, and H is the rotations of the square. Then the two cosets of H are $\{$ *elt*[1], *elt*[2], *elt*[3], *elt*[4] $\}$ = H = $H \times id$, and $\{$ *elt*[5], *elt*[6], *elt*[7], *elt*[8] $\}$ = $H \times$ *elt*[5]. A set of coset representatives is $\{$ elt[1], elt[5] $\}$ and the reader can verify that each element is uniquely expressible as a product of an element of H and a coset representative.

Conjugates and Normal Subgroups

A *conjugate* under G of an element h is an element of the form $g \times h \times g^{-1}$, for some element g of G. A conjugate under G of a subset H is a set of the form $g \times H \times g^{-1} = \{ g \times h \times g^{-1} \mid h \in H \}$, for some element g of G. The conjugate of a subgroup is again a subgroup. If S generates H then $g \times S \times g^{-1}$ generates $g \times H \times g^{-1}$.

For example, in the symmetric group of degree 4, let h=(1,2) and g=(1,3)(2,4). Then (3,4)=$g \times h \times g^{-1}$ is a conjugate of h. Let H=$\{$ (1,2), (3,4) $\}$. Then $g \times H \times g^{-1} = H$. Let $H = <$ (1,2), (3,4) $>$. Then $g \times H \times g^{-1} = H$ while $(2,4) \times H \times (2,4)^{-1} = <$ (1,4), (2,3) $>$.

An element g in G *normalizes* a subgroup H if the conjugate $g \times H \times g^{-1}$ is H. The set of all such elements is a subgroup, $N_G(H)$, called the *normalizer* of H in G.

For example, g=(1,3,2,4) normalizes $H = <$ (1,2), (3,4) $>$ in the symmetric group of degree 4. Indeed, $N_G(H) = < H, g >$.

A subgroup H whose normalizer in G is the whole group G is said to be a *normal* subgroup of G. Therefore all conjugates of a normal subgroup H are H itself. An alternative, but equivalent, definition of a normal subgroup is that

$$H \times g = g \times H, \text{ for } all\ g \in G.$$

The coset representatives of a normal subgroup can be chosen to have a special form. If $G = < H, s_1, s_2, \ldots, s_m >$ then the coset representatives can be chosen from $< s_1, s_2 \ldots, s_m >$. If $m = 1$, then the coset representatives are

$$id,\ s_1,\ s_1^2,\ \ldots,\ s_1^{n-1},$$

where s_1^n is the first power of s_1 that lies in H.

For example, the subgroup $H = \langle (1,2)(3,4), (1,3)(2,4) \rangle$ is normal in the symmetric group of degree 4. The group is generated by H together with s_1=(1,2,3) and s_2=(1,2). The coset representatives may be taken to be

$$id, s_1, s_1^2, s_2, s_1 \times s_2, s_1^2 \times s_2.$$

The group $\langle H, s_1 \rangle$ is also normal in the group. Its coset representatives may be taken to be

$$id, s_2$$

since $s_2^2 = id \in \langle H, g \rangle$. If instead we had s_2=(1,2,3,4) then the coset representatives may be taken to be

$$id, s_2$$

since $s_2^2 = (1,3)(2,4) \in \langle H, g \rangle$.

The cosets of a normal subgroup H in G form a group called the *factor group* or *quotient group* G/H. The inverse of a coset $H \times g$ is $H \times g^{-1}$, while the product of two cosets $H \times g_1$ and $H \times g_2$ is the coset $H \times (g_1 \times g_2)$. The inverse and product are independent of the choice of coset representative, because the subgroup H is normal.

Commuting Elements

Two elements g and h *commute* if $g \times h = h \times g$. For example, in these circumstances, we also say that g *centralizes* h, since $g \times h = h \times g$ implies that $g \times h \times g^{-1} = h$. The set of all elements that centralize h is a subgroup, $C_G(h)$, called the *centralizer* of h in G.

In the symmetric group of degree 4, the element (1,2) centralizes the element (3,4). The centralizer of (1,2) is just $\langle (1,2), (3,4) \rangle$.

Exercises

(1/Easy) If H is a subgroup of G and g, g' are elements of G, then show that $(H \times g) \times g'$ is a coset of H with coset representative $g \times g'$.

Bibliographical Remarks

There are many excellent texts on group theory and permutation groups. Two introductory texts are C.D.H. Cooper, **Permutations and Groups**, John Murray, London, 1975 and W. Ledermann, **Introduction to Group Theory**, Oliver-Boyd, Edinburgh, 1973. The standard reference is H. Wielandt, **Finite Permutation Groups**, Academic Press, New York, 1964.

A standard reference on group theory is M. Hall, Jr, **The Theory of Groups**, Macmillan, New York, 1959. An advanced text is I. D. Macdonald, **The Theory of Groups**, Oxford University Press, Oxford, 1968.

Chapter 3. List of Elements

This chapter develops Dimino's algorithm for constructing a list of elements of a group given by a set of generators. At each stage in the development, we will analyse the algorithms, and indicate the knowledge that is used by the algorithm.

Obvious Approach from First Principles

Suppose we are given a set of generators S of a group G. How do we determine a list of the elements of G, presuming this list is small enough to store in our computer? To answer this, let us consider what we know about groups and generators. We know that

(a) the group has an identity element;
(b) the set of generators S is a subset of the group G;
(c) a group is closed under multiplication, so that if we know two elements g and h in G then we must ensure their product $g \times h$ is in G; and
(d) the group G is the smallest set containing S which is closed under multiplication.

This suggests Algorithm 1.

Algorithm 1 : Closing under multiplication of elements

Input : a set S of generators of a group G;

Output : a list of the elements of the group G;

```
begin
  list := id ∪ elements of S;
  for each pair (g, h) of elements in list do
    if g × h is not in list then
      append g × h to list;
    end if;
  end for;
end;
```

To analyse the algorithm, there are $|G|^2$ pairs of elements (g, h), and for each pair we form one multiplication and perform one search of the list. The cost of appending an element to the list is insignificant, and occurs only $|G|$ times. Thus the total cost is

$|G|^2$ multiplications, and

$|G|^2$ searches of the list of elements.

The order of the group - and the length of the list - may be as large as 10 000. Therefore the searches must be performed as efficiently as possible. Today's implementations use a hash list of elements, so the average cost of a search is between 1.2 and 2 comparisons of elements. For permutations, we can assume that a comparison is equal in cost to a multiplication. In our analysis, we will assume the cost of a search is the same cost as two multiplications.

A Trivial Improvement

The algorithm may be improved slightly by observing one extra piece of information, namely

(e) for all elements g, $g \times id = id \times g = g$.

Hence, we should not consider the identity element as one of the pair of elements when we are closing the list of elements under multiplication. Alas, on analysis, we only reduce the $|G|^2$ factor to $(|G|-1)^2$.

Although this is not a major improvement, it does illustrate how additional knowledge can lead to improvement of algorithms. Our next example is more worthwhile.

Induction on the Length of Products

The next piece of knowledge we will draw upon comes from the definition of a set of generators. So far, we have only used the fact that the generators form a subset, and from them closure determines all the elements. The definition, however, says more. It says that

(f) each element of the group can be expressed as a product

$$s_{i_1} \times s_{i_2} \times \cdots \times s_{i_m}$$

of generators.

If we decompose the product into the first m-1 terms and the last term, then we may proceed by induction on the length of the product. Since (f) says that every element expressible as a product of length m is a product $g \times s$ of a generator s and an element g expressible as a product of length m-1, we modify our previous algorithm to only consider pairs of elements (g, h) where the second element is a generator. The new algorithm is Algorithm 2.

Algorithm 2 : Closing under multiplication with generators

Input : a set S of generators of a group G;

Output : a list of elements of the group G;

```
begin
  list := {id} ∪ elements of S;
  for each element g in list do
    for each generator s do
      if g × s is not in list then
        append g × s to list;
      end if;
    end for;
  end for;
end;
```

The analysis reveals that the algorithm requires

$|G| \times |S|$ multiplications, and

$|G| \times |S|$ searches.

If we work through the example of the symmetries of the square generated by $\{elt[2], elt[5]\}$ we obtain the following list of elements. We note the product that first indicated that the element is in the group.

elt [1], identity
elt [2], generator
elt [5], generator
$elt[3] = elt[2] \times elt[2]$
$elt[6] = elt[2] \times elt[5]$
$elt[8] = elt[5] \times elt[2]$
$elt[4] = elt[3] \times elt[2]$
$elt[7] = elt[3] \times elt[5]$

The number of multiplications performed is $16 = 8 \times 2$, as against 64 for Algorithm 1.

One Generator Groups

An important special case of the above algorithm occurs when there is only one generator. In this case we know the elements of the group are the powers of the generator s. This allows us to avoid searches in this case, by performing comparisons with the identity instead. One generator groups occur infrequently, so it would not pay to make any modifications to the above algorithm, if it was our general purpose algorithm for constructing lists of elements. However, one generator groups always play a role in Dimino's algorithm, and this special case is the best we can do for one generator groups.

One generator groups are usually called *cyclic* groups.

Induction on the Generators

Algorithm 2 formed the elements of the group, expressible as products

$$s_{i_1} \times s_{i_2} \times \cdots \times s_{i_m}$$

of the generators, in order of increasing length of the product. Dimino's algorithm forms the elements in order of the generators involved in the product. Thus, if $S = \{s_1, s_2, \cdots, s_t\}$ is the set of generators, all the elements involving the first i-1 generators are formed before any element that must involve the i-th generator s_i. The usefulness of this approach stems from the fact that the elements that involve only the first i generators form a subgroup. Now we can use our knowledge of subgroups and, in particular, cosets!

Some notation is required. Let $S_i = \{s_1, s_2, \cdots, s_i\}$, and let $H_i = \langle S_i \rangle$. For convenience, we let S_0 be the empty set, and H_0 be the identity subgroup. The typical inductive step of Dimino's algorithm is, given a list of elements for H_{i-1}, determine the list of elements for H_i.

Our knowledge of subgroups includes the fact that

(g) a subgroup is a group, and therefore is closed under multiplication.

In terms of the inductive step, this means that we know the product $g \times s$ is in the list, for all g in H_{i-1} and s in S_{i-1}. We could reorganise Algorithm 2 slightly so that, during the step from H_{i-1} to H_i, only the products $g \times s_i$, for g in H_{i-1}, and $g \times s$, for g not in H_{i-1} and s in S_i are considered, but this reorganisation would not save us anything over Algorithm 2.

The important facts about cosets are that

(h) the cosets of a subgroup partition a group, and

(i) all cosets have the same size as the subgroup. A coset can be formed by multiplying the elements of the subgroup by a coset representative.

Thus, if we know the elements of H_{i-1}, then whenever an element g is added to the list of elements for H_i we can in fact add the whole coset $H_{i-1} \times g$ to the list. If we extend the list of elements one coset at a time, then we know that all the elements of the coset $H_{i-1} \times g$ are not in the list once we know g is not in the list. This opens up the possibility of adding elements to the list while avoiding a lot of searching. However, the searches in Algorithm 2 come about in considering products $g \times s$ and determining if they are new elements. If we add cosets rather than elements we are still left with considering products $g \times s$ to find new coset representatives, so we have not avoided any searching. For the addition of cosets to be useful we need a better method of finding coset representatives than considering all products of elements and generators.

A new fact about cosets is needed. We have not met this fact before but it is easy to prove.

Lemma

Let H be a subgroup of G. Let $H \times g$ be a coset of H in G and let $s \in G$. Then all the elements of $(H \times g) \times s$ lie in the coset of H with representative $g \times s$.

The immediate consequence of this lemma is that we can find the coset representatives of H_{i-1} by considering the products $g \times s$ where s is in S_i and g is a previously found coset representative, rather than an arbitrary element. This is a major reduction in the number of products to be considered.

Taking the identity to be the coset representative of H_{i-1} gives Algorithm 3 for the inductive step of Dimino's algorithm.

Algorithm 3 : Inductive Step of Dimino's Algorithm

Input : a set S of generators of a group G;
a list of elements of H_{i-1};

Output : a list of elements of H_i;
begin
coset_reps := { *id* };
for each g in *coset_reps* **do**
for each generator s in S_i **do**
if $g \times s$ is not in *list* **then**
append $g \times s$ to *coset_reps*;
append $H_{i-1} \times (g \times s)$ to *list*;
end if;
end for;
end for;
end;

To analyse Algorithm 3, we note that there are i generators in S_i; there are $|H_i{:}H_{i-1}|$ cosets representatives; and there are $|H_i| - |H_{i-1}|$ elements added to the list. Thus there are $i \times |H_i{:}H_{i-1}|$ products formed when finding coset representatives, and $|H_i| - |H_{i-1}|$ products formed when appending elements to the list. Hence, there are a total of

$i \times |H_i{:}H_{i-1}| + (|H_i| - |H_{i-1}|)$ multiplications, and

$i \times |H_i{:}H_{i-1}|$ searches.

If we are careful about forming the elements of the coset then we do not form the coset representative twice. That is, when appending the coset to the end of the list, do not form the product of the identity and the coset representative. This reduces the number of multiplications by $|H_i{:}H_{i-1}| - 1$. Furthermore, only the product $id \times s_i$, that is s_i, has to be considered for the initial coset representative id. This reduces the number of multiplications by i.

Let us work through the second step for the example where G is the symmetries of the square generated by $\{elt[2], elt[5]\}$. Initially the list is

$elt[1]\ elt[2]\ elt[3]\ elt[4]$

and the coset representatives are

$elt[1]$

The generator $s_2 = elt[5]$ is not in the list, so it is the next coset representative. This appends the remainder of the elements to the list. The algorithm forms the products $elt[5] \times elt[2]$ and $elt[5] \times elt[5]$ and verifies that they are in the list. The algorithm terminates.

Let us consider the same group, but with generators $\{elt[5], elt[2]\}$. Again we perform the second inductive step. The list initially is *elt* [1] *elt* [5]

and the coset representatives are

elt [1]

The generator *elt* [2] is the next coset representative, and its coset is added to the list. Now the list is

elt [1] *elt* [5] *elt* [2] *elt* [6]

and the coset representatives are

elt [1] *elt* [2]

The product *elt* [2] × *elt* [5] is *elt* [6], which is in the list. The product *elt* [2] × *elt* [2] is *elt* [3], which is the next coset representative. Its coset is added to the list giving

elt [1] *elt* [5] *elt* [2] *elt* [6] *elt* [3] *elt* [7]

and the coset representatives are

elt [1] *elt* [2] *elt* [3]

The product *elt* [3] × *elt* [5] is *elt* [7], which is in the list. The product *elt* [3] × *elt* [2] is *elt* [4], which is the next coset representative. Its coset is added to the list completing the group. The products *elt* [4] × *elt* [5] and *elt* [4] × *elt* [2] are formed and found to be in the list. The algorithm terminates.

Simple Algorithm of Dimino

We now incorporate the inductive step with a few modifications to obtain the simple version of Dimino's algorithm. The first inductive step involves a one generator group, so we treat this as a special case and do not use Algorithm 3 at this step. Instead we use the modified Algorithm 2 discussed in the previous section. If the first element in the list is the identity then the first element of a coset will always be the coset representative. Each coset has size $|H_{i-1}|$ during the *i*-th step. Hence, we can easily locate the coset representatives in the list, and so we do not store them separately. We also incorporate a check on the redundancy of the generators. There is no inductive step if a generator is already in the list. The simple algorithm of Dimino is Algorithm 4.

Algorithm 4 : Simple Dimino's Algorithm

Input : a set $S = \{s_1, s_2, \ldots, s_t\}$ of generators of a group G;

Output : a list of elements of the group G;

```
begin
  (*Treat the special case <s_1>*)
  order := 1; elements[1] := id;
  g := s_1;
  while not (g = id) do
    order := order + 1; element[order] := g;
    g := g × s_1;
  end while;

  (*Treat remaining inductive levels*)
  for i := 2 to t do
    if not s_i in elements then (*next generator is not redundant*)
      previous_order := order; (*i.e. |H_{i-1}|*)
      (*first useful coset representative is s_i - add a coset*)
      order := order + 1; elements[order] := s_i;
      for j := 2 to previous_order do
        order := order + 1; elements[order] := elements[j] × s_i;
      end for;

      (*get coset representative's position*)
      rep_pos := previous_order + 1;
      repeat
        for each s in S_i do
          elt := elements[rep_pos] × s;
          if not elt in elements then (*add coset*)
            order := order + 1; elements[order] := elt;
            for j := 2 to previous_order do
              order := order + 1; elements[order] := elements[j] × elt;
            end for;
          end if;
        end for;
        (*get position of next coset representative*)
        rep_pos := rep_pos + previous_order;
      until rep_pos > order;

    end if; (* not s_i ... *)
  end for; (* i := 2 ... *)
end;
```

Let $N_i = |H_i{:}H_{i-1}|$, so that $N_1 = |H_1|$ and $|G| = \prod_{i=1}^{t} N_i$. If we assume there are no redundant generators (and if we omit the t-1 searches associated with the redundancy checking), then the total number of multiplications for Algorithm 4 is N_1 for the initial special case plus the sum of the inductive steps. The inductive steps are careful to avoid the multiplications mentioned in the previous section, so the total number of multiplications is

$$N_1 + \sum_{i=2}^{t} \left[i \times (N_i - 1) + (N_i - 1) \times (|H_{i-1}| - 1) \right]$$
$$= N_1 + \sum_{i=2}^{t} \left[(i-1) \times (N_i - 1) \right] + \sum_{i=2}^{t} \left[|H_i| - |H_{i-1}| \right].$$

The last summation evaluates to $|G| - |H_1|$, so the total is

$$|G| + \sum_{i=2}^{t} \left[(i-1) \times (N_i - 1) \right].$$

For the common case, where $t = 2$, the number of multiplications is maximised when $N_1 = 2$, and $N_2 = |G|/2$. The maximum is $1.5 \times |G| - 2$ multiplications.

An analysis of the number of searches gives N_1 comparisons with the identity, for the special case, plus $\sum_{i=2}^{t} \left[i \times (N_i - 1) \right]$ searches for the inductive steps. This is maximised as above for the case when $t = 2$ to be approximately $|G|$ searches.

As an example, let us consider the symmetric group on $\{1,2,3,4\}$. The group has order 24. The generators we choose are (1,2), (3,4), (1,3,2,4), and (2,3,4). This gives values of $N_1 = 2$, $N_2 = 2$, $N_3 = 2$, and $N_4 = 3$. The subgroups form a chain of subgroups

$$id < H_1 < H_2 < H_3 < H_4 = G$$

which provides the basis of the induction. The aim of Dimino's algorithm is to make the cost dependent on the sum of the sizes of the steps from H_{i-1} to H_i - that is, on $\sum_{i=1}^{4} N_i$ - instead on the total size of the group - that is, on $\prod_{i=1}^{4} N_i$. The cost of Algorithm 2 for this example is 96 multiplications and 96 searches, which is equivalent to a total of 288 multiplications. The cost of Algorithm 4 is 33 multiplications, 2 comparisons, and 13 searches, which is equivalent to a total of 61 multiplications.

Complete Algorithm of Dimino

There is still a minor improvement that can be made to the algorithm. However, we need the concept of a normal subgroup. A subgroup H of G is *normal* in G if $H \times g = g \times H$, for all elements g of G. A consequence of this is that if $G = \langle H, g \rangle$ then the coset representatives of H in G can be taken to be

$$id,\ g,\ g^2, \ldots, g^m$$

where m is the smallest integer such that $g^{m+1} \in H$. We can test if H is normal in $\langle H, g \rangle$ by testing if all the elements $g^{-1} \times s \times g$, for s a generator of H, are in H. If so, then H is normal.

The cost of testing if H_{i-1} is normal in H_i is $2 \times (i\text{-}1)$ multiplications to form the elements above, and $(i-1)$ searches to test if they are in the subgroup. The cost of the inductive step if H_{i-1} is normal in H_i is $(N_i - 1) \times (|H_{i-1}| - 1)$ multiplications to add the cosets, $N_i - 1$ multiplications to form the powers of s_i as coset representatives, and $N_i - 1$ searches to see if the powers are in the subgroup. This saves $(i-1) \times (N_i - 1)$ multiplications and $(i-1) \times (N_i - 1)$ searches over the typical inductive step (ignoring the cost of normality testing).

For the example of the symmetric group on {1,2,3,4} given by four generators, the complete Dimino's algorithm works out to be more expensive. It requires 12 multiplications and 6 searches while testing normality. The first two inductive steps do involve normal subgroups. Their cost (outside of normality testing) is 6 multiplications and 2 searches. The initial special case for the cyclic group, and the last step cost 24 multiplications, 2 comparisons, and 8 searches, outside of normality testing. Hence, the total cost is equivalent to 76 multiplications.

Uses

A list of elements allows us to answer two vital questions about a group.

(1) What is the order of the group?

(2) Is a given permutation in the group?

A list of elements defines a bijection between the group and the integers from 1 to $|G|$. Representing an element by its position in the list is more space efficient and often more convenient than representing it as a permutation.

Using the bijection, a subset or subgroup of a group can be represented as a characteristic function (or bitstring) of length $|G|$ bits. The i-th bit is set if and only if the i-th element is in the subset.

The list of elements provides a simple means of running through the elements of the group without repetition. This is useful in many other algorithms, particularly those involving searches of the group.

Summary

Obtaining a list of elements is the first step in studying a small group, and Dimino's algorithm is the best known algorithm for obtaining the list. The summation terms in the analysis of Dimino's algorithm can be crudely bounded by the order of the group. Therefore, the total cost is never more than $2 \times |G|$ multiplications and $|G|$ searches.

Exercises

(1/Easy) For the small examples of Chapter 2, choose a suitable set of generators and construct a list of elements.

(2/Easy) The symmetries of the projective plane of order two may be generated by a=(4,6)(5,7), b=(4,5)(6,7), c=(2,6,4)(3,7,5), d=(2,4)(3,5), e=(1,2,4,5,7,3,6). The relevant indices are N_1=2, N_2=2, N_3=3, N_4=2, N_5=7. Noting that all steps except the last have H_{i-1} normal in H_i, what is the cost of the complete Dimino's algorithm?

How does it compare with the cost of the simple Dimino's algorithm?

Conclude that the index of a normal step must be greater than or equal to three for there to be any gain in using the complete algorithm.

Bibliographical Remarks

The earlier algorithms in this chapter have been suggested at various times in the literature. For example, an algorithm that multiplies every pair of elements is presented in Yu P. Vasil'ev, "*One method of computer-aided description of the lattice of subgroups of a finite group*", Kibernetika **2** (1978) 131-133. There is also an algorithm that was the state of the art until Dimino's algorithm. It is due to Joachim Neubüser, "*Untersuchungen des Untergruppenverbandes endlicher Gruppen auf einer programmgesteuerten elektronisch Dualmaschine*", Numerische Mathematik **2** (1960) 280-292.

Dimino's algorithm was developed by Lou Dimino of Bell Laboratories around 1970. He did not publish his algorithm, but through John Cannon it found its way into the GROUP system and into the folklore of the field. The first description in the literature is in Harald Steinmann, **Ein transportables Programm zur Bestimmung des Untergruppenverbandes von endlichen auflösbaren Gruppen**, Diplomarbeit, Aachen, 1974.

Details of hash searching, including the average number of comparisons, can be found in D. E. Knuth, **The Art of Computer Programming, volume 3 : Sorting and Searching**, Addison-Wesley, Reading, Massachusetts, 1973.

A variation on Dimino's algorithm to compute the list of elements in a double coset $H \times g \times K$, appears in G. Butler, "*Double cosets and searching small groups*", **SYMSAC'81** (Proceedings of the 1981 ACM Symposium on Symbolic and Algebraic Computation, Snowbird, Utah, August 5-7, 1981) (P. Wang, ed.), ACM, 1981, 182-187.

Chapter 4. Searching Small Groups

This chapter develops algorithms for searching small groups. Many algorithms in computational group theory involve searching a group for an element (or elements) with specific properties. We will look at the general principles involved in these searches, and illustrate them by finding elements that commute with (or centralize) a given element. While we consider these principles in the context of small groups, they do apply to large permutation groups as well.

We will consider elements that satisfy a given property P. Of course, we must be able to determine whether an element satisfies the property or not. If the element g satisfies the property P, we write $P(g)$.

We begin with algorithms that find one element with the specified property. Then we consider algorithms for finding all the elements with a specified property, in the special case where the set of all such elements forms a subgroup.

Linear Search of the List of Elements

Suppose we are searching a group G for an element that satisfies a given property P. For small groups we have a list of elements, so our search could just run through this list. This is the approach taken in Algorithm 1.

Algorithm 1 : Search List of Elements for One Element

```
Input : a list of elements of a group G;
      a property P;

Output : an element g with property P, if one exists;

begin
  for each element g of G do
    if P(g) then
      exit with result( g );
    end if;
  end for;
end;
```

Let $c(P)$ be the (average) cost of determining whether an element satisfies property P. Then the worst case cost of Algorithm 1 is

$$|G| \times c(P).$$

For a concrete example, let z be a fixed element of the group G. Let the property P be

$$P : g \in G,\ g \notin \langle z \rangle,\ g \times z = z \times g.$$

The elements that satisfy P are those that commute with (or centralize) z. Assuming that no elements in the cyclic group $<z>$ are actually tested, the cost of determining the property is

2 multiplications, and

1 comparison of elements.

Searching as a Discarding Process

Let us view the linear search of the list of elements in another way. This alternate view allows us to generalize the algorithm, and improve the search. Instead of running through the list of elements, let us regard the search as a process for decreasing a set of possible solutions by discarding non-solutions. Initially, the whole group G will be the set of possible solutions. The process will end when one element in the set is found that has the specified property. With this view, Algorithm 1 for finding an element that commutes with z is transformed into Algortihm 2.

Algorithm 2 : Discarding elements to find one element

```
Input : a list of elements for a group G;
    a property P;

Output : an element g with property P, if one exists;

begin
  Γ := G - <z>;
  while Γ is not empty do
    choose g ∈ Γ;
    if P(g) then
      exit with result(g);
    else
      Γ := Γ - {g};
    end if;
  end while;
end;
```

As noted in Chapter 2, an effective way to represent the subset Γ is as a bitstring of length $|G|$.

One approach to improving the search algorithm is to discard more than one element of the group at a time. This attempts to avoid tests for the property P. Hopefully, the cost of discarding the extra elements will be less than the cost of the tests avoided.

Just how much can we discard? Let $Ded_P(g)$ be a set of elements y of G for which we can "easily deduce" not $P(y)$ from not $P(g)$. By this we mean that the set can be constructed from the property P and the element g, without global knowledge of the group G, (that is, without running through the list of elements of G). For example, from not $P(g)$ in the case of commuting elements, we can easily deduce

not $P(g^{-1})$,
not $P(z \times g)$,
not $P(g \times z)$,
not $P(z^m \times g)$, for any integer m,
not $P(g \times z^m)$, for any integer m, and
not $P(z^m \times g \times z^n)$, for any integers n and m.

but we could not easily deduce

not $P(y)$, for $y^m = g$, for some integer m

even though it is deducible from not $P(g)$, since the only practical way of finding such y's would be to run through the group G taking powers.

The set $Ded_P(g)$ is in some sense the largest set of elements we could discard from the set of possible solutions. In practice, we require a subset $D_P(g)$ of $Ded_P(g)$ that is cheap to compute - at least more cheaply than performing the tests. Typically a coset of some subgroup is used as the subset. For our particular property P, one could use $D_P(g) = \langle z \rangle \times g$ as the discard subset. The cost of forming a coset is one multiplication per element of the coset (and in this instance, one search per element to create the subset as a bitstring). This is the same as the cost of one test per element, (under our usual assumptions about the cost of a search).

The search algorithm is now Algorithm 3.

Algorithm 3 : Discarding cosets to find one element

Input : a list of elements of a group G;
    a property P;

Output : an element g with property P, if one exists;

begin
  $\Gamma := G - \langle z \rangle$;
  while Γ is not empty **do**
    choose $g \in \Gamma$;
    if $P(g)$ **then**
      exit with result(g);
    else
      $D_P(g) := \langle z \rangle \times g$; $\Gamma := \Gamma - D_P(g)$;
    end if;
  end while;
end;

An analysis of the worst case of Algorithm 3 follows. The worst case occurs when no element satisfies the property. In this case, all the elements of the group are eventually discarded from the set of possible solutions. The cost of forming $D_P(g)$ as a bitstring is $|z|-1$ multiplications and searches. That is, one multiplication and search for each element in the set that was not tested. The number of elements tested is $|G|/|z|$, with the remaining elements being discarded as part of $D_P(g) - \{g\}$, for some g. The cost of each test is one comparison and two multiplications. Thus the worst case cost is

$$\frac{|G|}{|z|}$$ comparisons,

$$\frac{|G|}{|z|} \times 2 + \left[|G| - \frac{|G|}{|z|} \right]$$ multiplications, and

$$|G| - \frac{|G|}{|z|}$$ searches.

Taking a comparison to be equivalent to a multiplication, and a search to be equivalent to two comparisons, gives a total worst case cost of

$3 \times |G|$ multiplications.

This is also the worst case cost of Algorithm 1, because for this example the cost of discarding an element is the same as performing a test of the property.

Another Example

An example where the cost of discarding an element is less than the cost of performing a test of the property occurs when one is searching for an element g that normalizes a given subgroup U. Let S be a set of generators for U. Then the property is

$$P : g \in G,\ g \notin U,\ g^{-1} \times s \times g \in U, \text{ for } each\ s \in S.$$

To test the property requires $|S| \times 2$ multiplications and $|S|$ searches. (This assumes the inverse of the element g is known, or that the cost of a "division" of permutations is the same as a multiplication.) We can take the discard set $D_P(g)$ to be the coset $U \times g$. Then the cost of discarding an element is one multiplication and one search.

In the worst case where no element normalizes U, there are $|G|/|U|$ tests, and the remaining elements are discarded. This gives a cost of

$$2 \times |S| \times \frac{|G|}{|U|} + \left[|G| - \frac{|G|}{|U|} \right]$$ multiplications, and

$$|S| \times \frac{|G|}{|U|} + \left[|G| - \frac{|G|}{|U|} \right]$$ searches.

This gives a total equivalent to

$$\frac{|G|}{|U|} \times \left[4 \times |S| + 3 \times |U| - 3 \right]$$

multiplications.

When the subgroup U has two generators, this represents a saving of $5 \times \left[1 - \frac{1}{|U|} \right] \times |G|$ multiplications over Algorithm 1, which requires $4 \times |S| \times |G|$ multiplications.

Finding All Elements

Often the purpose of a search is to find all the elements with a specific property knowing that these elements form a subgroup. For example, the set of elements that commute with a given element z forms a subgroup called the centralizer of z in G. In this section, we look at adapting our search algorithm for one element to find all elements.

The search for a single element attempts to discard as many elements as possible from the set of possible solutions. However, it only does this for elements it knows do not have the specified property. Once it finds an element with the property it stops. The search for all elements has the possibility of discarding elements that it has already included in the subgroup. Thus the set of possible solutions is reduced in two ways.

Let H be the set of elements found so far to have property P. The set will be a subgroup. Let $D_P(g)$ be the discard set for the element g that does not satisfy P. Then our algorithm is Algorithm 4.

Algorithm 4 : Subgroup of elements with a given property

Input : a list of elements of a group G;
    a property P such that the elements satisfying P form a subgroup;

Output : the subgroup H of G of elements that satisfy the property;

```
begin
  H := { id }; Γ := G - H;
  while Γ is not empty do
    choose g ∈ Γ;
    if P(g) then (* add to H those with the property *)
      H := < H, g >;  Γ := Γ - H;
    else (* discard others without the property *)
      Γ := Γ - D_P(g);
    end if;
  end while;
end;
```

Since the set of all elements satisfying the property P forms a subgroup, it follows from $P(g)$ that all elements of $< H, g >$ satisfy P. Hence they are added to H and discarded from Γ. Forming the new subgroup H is just the inductive step of Dimino's algorithm modified to work with bitstrings.

Furthermore, we may take, for any such property P, the coset $H \times g$ as the discard set $D_P(g)$. This is so because all the elements of H satisfy P while g does not satisfy P.

Analysis

The analysis of Algorithm 4 is not straightforward, and is still incomplete. We will cover the initial ground work in this section to provide a crude upper bound on the cost. In the next section we provide a more detailed analysis.

The algorithm, in essence, finds generators

$$id = s_0, s_1, s_2, \ldots, s_t$$

for the final subgroup H. The values taken by the variable H in the algorithm are $H_i = \langle s_1, s_2, \ldots, s_i \rangle$, for i from 0 to t. Let us first consider the number of tests performed during the algorithm. Suppose the elements tested were the sequence $\{g_i\}$, and that i_k elements were tested between the finding of s_{k-1} and s_k. For convenience, let g_0 be the identity, and let i_{t+1} be one more than the number of tests performed after finding the last generator. Then

$$g_{i_1} = s_1,$$
$$g_{i_1 + i_2} = s_2,$$
$$\vdots$$
$$g\left(\sum_{j=1}^{k} i_j\right) = s_k,$$
$$\vdots$$

While the value of H is H_k we form $i_{k+1} - 1$ discard sets, and perform at most one inductive step of Dimino's algorithm. If $D_P(g)$ is the coset $H \times g$ then the cost of forming the discard set is one multiplication and one search per element (different from g). Let the constant c denote this cost per element. Then the total cost of the discards is

$$\sum_{k=0}^{t} \left[c \times \left(i_{k+1} - 1 \right) \times |H_k| \right].$$

The total effort of forming the subgroups is equivalent to the complete Dimino's algorithm for $H = \langle s_1, s_2, \ldots, s_t \rangle$. This cost was analysed in the third chapter. The adjustments to that analysis due to the use of bitstrings are minor. It is just the additional cost of one search per element of the final subgroup H_t in order to obtain their positions in the list of elements of G.

Let us obtain an upper bound for the values i_k. This will give us an upper bound on the cost of the algorithm.

The value of $i_{t+1} - 1$ is bounded by the number of cosets of H_t in G minus one, since each discard set is a coset of H_t distinct from H_t.

For an upper bound on i_k, recall that the value of H will change once we find any element of the final subgroup H_t. There are $|G|/|H_{k-1}|$ cosets of H_{k-1} that could possibly be considered for discarding, but of these $|H_t|/|H_{k-1}|$ give an element that satisfies the property. Thus $i_k - 1$, the number of cosets discarded, is bounded by

$$\frac{|G|-|H_t|}{|H_{k-1}|}.$$

Thus

$$c \times \left[i_{k+1} - 1 \right] \times |H_k| \quad \leq \quad c \times \left[|G| - |H_t| \right]$$

and the cost of discarding is bounded by

$$c \times (t+1) \times \left[|G| - |H_t| \right].$$

Multiple Discarding

The above upper bound assumes that the values of i_k are independent of each other. This is not the case, so we should expect to be able to refine the bound.

The above upper bound says that each element not in the subgroup is discarded the maximum number of times; that being $(t+1)$. While this figure is not exact, multiple discarding of elements does occur. There are $|H_k:H_{k-1}|$ cosets of H_{k-1} in one coset of H_k, and many of these could be discarded while the value of the variable H is H_{k-1} without all of them being discarded. Therefore, there is the possibility of locating an element that causes the coset of H_k to be discarded. This may discard some cosets of H_{k-1} again. Hence the occurrence of multiple discarding.

There are $|G|-|H_t|$ elements without the property P, and $(i_k - 1) \times |H_{k-1}|$ of these have been discarded in cosets of H_{k-1}. Thus there are

$$|G|-|H_t| - \left[i_k - 1 \right] \times |H_{k-1}|$$

elements without the property that could be selected as the element g in the algorithm.

There are

$$\frac{|G|-|H_t|}{|H_k|}$$

cosets of H_k containing elements without the property. It only requires one element of a coset to be still in the set Γ and the whole coset could be discarded. So we are only guaranteed that a complete coset of H_k has been discarded as a union of cosets of H_{k-1} if the number of elements still in Γ is less than the number of cosets of H_k without the property P. Hence, the number of complete cosets of H_k discarded as cosets of H_{k-1} is at least

$$M = \max \left[0, \frac{|G|-|H_t|}{|H_k|} - \left[|G|-|H_t| - \left[i_{k-1} - 1 \right] \times |H_{k-1}| \right] \right].$$

Thus $i_{k+1} - 1$ is bounded by

$$\frac{|G|-|H_t|}{|H_k|} - M \;=\; \min \left[\frac{|G|-|H_t|}{|H_k|}, |G|-|H_t| - \left[i_k - 1 \right] \times |H_{k-1}| \right].$$

Let

$$j_k = c \times \left[i_{k+1} - 1 \right] \times |H_k|, \text{ and}$$
$$K = c \times \left[|G| - |H_t| \right]$$

then, for the worst case analysis, we wish to maximise $\sum_{k=0}^{t} j_k$ subject to the constraints

$$j_{-1} = 0, \text{ and}$$

$$0 \le j_k \le \min \left[K, \left[K - j_{k-1} \right] \times |H_k| \right], k = 0, 1, \ldots, t.$$

A solution to this problem has not been found to date.

Summary

Searching for subgroups and elements is an integral part of many other algorithms, as well as being essential in the study of groups. Viewing it as a discarding process is a good paradigm for improvements, and for analogy with search algorithms for large permutation groups.

Typically the search algorithms discard cosets (of some subgroup). The analysis of these algorithms is not straightforward and multiple discarding is not yet fully analysed. Other choices of discard sets have rarely been investigated, and their analysis is more incomplete.

Exercises

(1/Easy) In the symmetric group of degree 4, execute Algorithm 3 to find a centralizing element of z=(1,2,3,4).

(2/Easy) In the symmetric group of degree 4, execute Algorithm 4 to construct the normalizer of $H = \langle (1,2,3,4) \rangle$. Use $D_P(g) = H \times g$. Do the same for $H = \langle (1,2)(3,4), (1,3)(2,4) \rangle$.

(3/Very Difficult) Extend the analysis of multiple discarding to account for all the previous values of H.

Bibliographical Remarks

The search algorithm was developed to compute the normalizer of a subgroup. This information was required by the lattice of subgroups algorithm (in Chapter 6). Algorithms to compute the normalizer in this context have been suggested and implemented by various people. The references are V. Felsch and J. Neubüser, "*Ein Programm zur Berechnung des Untergruppenverbandes einer endlichen Gruppe*", Mitteilungen des Rheinish-Westfälischen Institutes für Instrumentelle Mathematik (Bonn) **2** (1963) 39-74; V. Felsch, **Ein Programm zur Berechnung des Untergruppenverbandes und der Automorphismengruppe einer endlichen Gruppe**, Diplomarbeit, Kiel, 1963; J. J. Cannon, **Computation in Finite Algebraic Structures**, Ph. D. Thesis, Sydney, 1969; and J.J. Cannon and J. Richardson, "*The GROUP system for investigating the structure of groups*", Technical Report 8, Computer-Aided Mathematics Project, Department of Pure Mathematics, University of Sydney, 1971. The Felsch algorithm is essentially the algorithm of this chapter. It differs in that it retains information about the representatives of the cosets discarded. The author first saw Felsch's algorithm towards the end of 1975 in the implementation of the subgroup lattice algorithm in

GROUP at that time. The exposition of this chapter is a simplified presentation of that algorithm.

Algorithm 4 is suggested in P. Dreyer, **Ein Programm zur Berechnung der auflösbaren Untergruppen von Permutationsgruppen**, Diplomarbeit, Kiel, 1970, however, Dreyer implemented the Felsch/Neubüser algorithm.

Another choice of discard set when searching for a subgroup, namely the double coset $H \times g \times H$, is investigated in G. Butler, "*Double cosets and searching small groups*", **SYMSAC'81** (Proceedings of 1981 ACM Symposium on Symbolic and Algebraic Computation, Snowbird, Utah, August 5-7, 1981), (P. Wang, ed.), ACM, 1981, 182-187. In general the cost of computing double cosets is too expensive. This paper is the initial source of our analysis. In particular, it introduces the study of multiple discarding.

Chapter 5. Cayley Graph and Defining Relations

This chapter looks at an interesting graphical representation of a small group. The representation is called the Cayley graph and may be considered to be an abbreviated multiplication table for the group. The representation is used to determine a set of defining relations for the group. These are the laws or axioms obeyed by the group's generators (beyond the axioms obeyed by all groups) that specify the particular group. The defining relations provide a connection between small groups and the combinatorial group algorithms.

Definitions

For a group given by a set of generators, a *Cayley graph* represents the effects of multiplying an element of a group by a generator of the group. For each element there is a node of the graph. From each node and for each generator there is a directed edge, labelled by the generator, that leads to the node corresponding to the product of the element and the generator.

For example, in the Cayley graph of the symmetries of the square, where the generators are $a = elt\,[2]$ and $b = elt\,[5]$, there are edges

$$elt[5] \xrightarrow[a]{\qquad\qquad} elt[8]$$

$$elt[5] \xrightarrow[b]{\qquad\qquad} elt[1]$$

since $elt\,[5] \times elt\,[2] = elt\,[8]$, and $elt\,[5] \times elt\,[5] = elt\,[1]$. The complete Cayley graph for this example is given in Figure 1. The nodes are integers corresponding to the elements $elt\,[1]$ to $elt\,[8]$.

A *word* in the generators is a product

$$s_{i_1} \times s_{i_2} \times \cdots \times s_{i_m}$$

where each term is a generator, or the inverse of a generator. Some examples are a, b, b^2, $a \times b$, a^6, and $a \times b \times a \times b \times a^{-1}$. A word is a formal (or symbolic) product. When a word is evaluated we get an element of the group.

Figure 1 : Cayley Graph of Symmetries of the Square

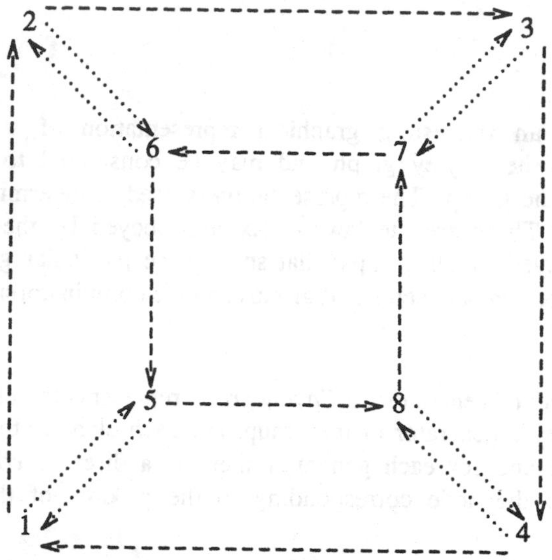

The dashed arrows represent edges labelled by *a*.
The dotted arrows represent edges labelled by *b*.

In the group of symmetries of the square,

a	evaluates to	*elt* [2],	
b	evaluates to	*elt* [5],	
b^2	evaluates to	*elt* [1],	
$a \times b$	evaluates to	*elt* [6],	
a^6	evaluates to	*elt* [3],	and
$a \times b \times a \times b \times a^{-1}$	evaluates to	*elt* [4].	

There are an infinite number of words that evaluate to any given element.

The inverse of a word is

$$s_{i_m}^{-1} \times s_{i_{m-1}}^{-1} \times \cdots \times s_{i_1}^{-1}.$$

It evaluates to the inverse of the element to which the word evaluates.

By tracing words through the Cayley graph we can multiply elements. To form *elt* [7] × *elt* [3] we require a word that evaluates to *elt* [3]. Such a word is a^6. Starting at node 7 we trace a path along edges whose labels correspond to the terms in the word. As shown in Figure 2, this leads to the product of *elt* [7] and *elt* [3] being *elt* [5].

Figure 2 : Tracing as Multiplication

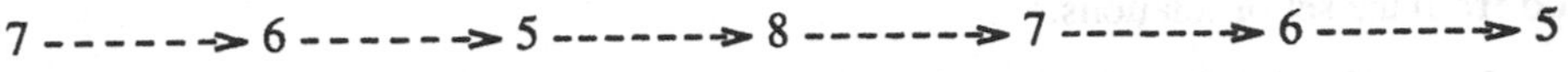

If the word has a term which is the inverse of a generator then follow the edge labelled with that generator, but in the reverse direction. Thus tracing $a^{-1} \times b \times a^{-1}$ from node 7 proceeds to node 3 as shown in Figure 3.

Figure 3 : Tracing Inverses

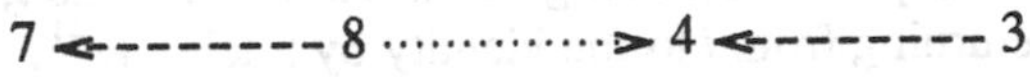

We can determine a word for *elt* [3] by finding a path from the identity (node 1) to node 3. To find a word for each element it is useful to have a spanning tree of the Cayley graph. One spanning tree for the above Cayley graph is given in Figure 4.

Figure 4 : Spanning Tree

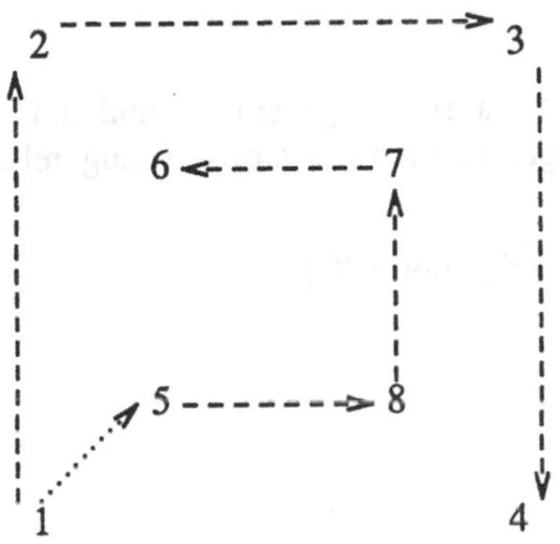

This gives a unique word for each element of the group. For this example the correspondence between elements and words is:

element 1	corresponds to	empty word,	
element 2	corresponds to	a,	
element 3	corresponds to	a^2,	
element 4	corresponds to	a^3,	
element 5	corresponds to	b,	
element 6	corresponds to	$b \times a^3$,	
element 7	corresponds to	$b \times a^2$,	and
element 8	corresponds to	$b \times a$.	

A *relation* is a word corresponding to the identity element. Some examples in this case are a^4, a^8, a^{12}, b^2, $a^4 \times b^2$, b^4, and $b \times a \times b \times a$. The relations corresponding to the group axioms are called *trivial relations*. They hold for every group. A relation corresponds to a loop in the Cayley graph that begins and ends at the identity node.

Note that the relations in the above example are not independent. A set of relations is called a set of *defining relations* for the group if every (nontrivial) relation for the group can be deduced from the set of relations.

We can always deduce the trivial relations from a set of relations. The set of deductions is closed under products, inverses, and conjugates of the form $w \times r \times w^{-1}$, for some word w and deduction r.

Viewing relations as loops about the identity, we see that a product of relations is just a concatenation of loops; an inverse of a relation is the loop in the reverse direction; and a conjugate $w \times r \times w^{-1}$ corresponds to tracing the word w to a node i, looping around i with the relation r, and then returning from i to the identity by tracing w^{-1}.

Start with a Spanning Tree

The number of loops around the identity is infinite, so taking all relations of a group and attempting to eliminate deductions is not a feasible approach for determining a set of defining relations. This section aims to reduce the number of loops/relations that must be considered. The role of the spanning tree is evident from the following theorem.

Loop Basis Theorem

Suppose a group G is defined by a set of generators and a Cayley graph. Given a spanning tree T of the Cayley graph, then a set of defining relations for the group is formed by the set

$$R(T) = \{ R(e) \mid e \text{ is an edge not in } T \}.$$

The relation $R(e)$ for the edge

$$e : i \overset{s}{\dashrightarrow} j$$

is the word $w_i \times s \times w_j^{-1}$, where w_i is the word/path in the spanning tree from node 1 to node i.

Consider the example of the symmetries of the square for which we already have a Cayley graph and a spanning tree. The set of defining relations $R(T)$ we obtain is

$R(2\,..\,b\,..>6)$	$\equiv$	$(a) \times b \times (a^{-3} \times b^{-1})$
$R(3\,..\,b\,..>7)$	$\equiv$	$(a^2) \times b \times (a^{-2} \times b^{-1})$
$R(4\,\text{--}\,a\,\text{--}>1)$	$\equiv$	$(a^3) \times a = a^4$
$R(4\,..\,b\,..>8)$	$\equiv$	$(a^3) \times b \times (a^{-1} \times b^{-1})$
$R(5\,..\,b\,..>1)$	$\equiv$	$(b) \times b = b^2$
$R(6\,..\,b\,..>2)$	$\equiv$	$(b \times a^3) \times b \times (a^{-1})$
$R(6\,\text{--}\,a\,\text{--}>5)$	$\equiv$	$(b \times a^3) \times a \times (b^{-1})$
$R(7\,..\,b\,..>3)$	$\equiv$	$(b \times a^2) \times b \times (a^{-2})$
$R(8\,..\,b\,..>4)$	$\equiv$	$(b \times a) \times b \times (a^{-3})$

Note that the relation $R(e)$ is defined even for edges in the spanning tree. In this case, the loop lies wholly within the spanning tree and the relation is a trivial relation.

To prove the theorem we must show that any relation/loop is deducible from the set $R(T)$. Take any loop

$$id = j_0 - s_{i_1} \rightarrow j_1 - s_{i_2} \rightarrow j_2 \cdots j_{m-1} - s_{i_m} \rightarrow j_m = id$$

about the identity, corresponding to the relation

$$s_{i_1} \times s_{i_2} \times \cdots \times s_{i_m} = identity.$$

At each node visited in the loop we can insert a path through the spanning tree to the identity node, and back again. Each of these is the insertion of a trivial relation. The resulting loop is the concatenation of the loops $R(j_{k-1} - s_{i_k} \rightarrow j_k)$. These are either in $R(T)$, or are themselves trivial relations. Hence, all relations are deducible from $R(T)$.

The size of $R(T)$ is the number of edges not in the spanning tree. If S is the set of generators, then the size of $R(T)$ is

$$1 + |G| \times \left[|S| - 1 \right].$$

This is still often larger than is necessary so we will look at ways of choosing a subset of $R(T)$ from which we can deduce the remaining relations in $R(T)$.

The Colouring Algorithm

The colouring algorithm determines a subset of $R(T)$ that is a set of defining relations. The idea behind the algorithm is to colour those edges e for which $R(e)$ can be deduced from the current set of relations. Initially the edges of the spanning tree are coloured because they correspond to the trivial relations. When a new relation of $R(T)$ is added to the subset then we determine all deductions from the subset and colour the appropriate edges. A relation of $R(T)$ is added only if it currently corresponds to an uncoloured edge.

The heart of the algorithm is determining the deductions. For a single relation r in the subset we can deduce $w_i \times r \times w_i^{-1}$ for all nodes i in the graph. This is the loop r around node i. If this loop around i contains precisely one uncoloured edge e then we can colour the edge. The argument for doing this is the same as in the proof of the Loop Basis Theorem. The loop about node i can be considered as a concatenation of loops $R(f)$. All the edges f (except e) are coloured, so the relations can be deduced from the subset. But the whole loop can also be deduced from the subset. Hence $R(e)$ can be expressed as a product involving $w_i \times r \times w_i^{-1}$ and the other $R(f)$ involved in the loop, all of which are deductions from the subset. Hence $R(e)$ is deducible from the subset.

To determine all the deductions from the subset we repeatedly trace single relations in the subset about the nodes of the graph colouring edges where possible until we can colour no more edges. The simplest strategy for determining the deductions is illustrated in Algorithm 1.

Algorithm 1 : Defining Relations by Colouring Cayley Graph

Input : a group G given by a set S of generators and a Cayley graph;

Output : a set of defining relations;

```
begin
  construct a spanning tree of the Cayley graph;
  colour the edges of the spanning tree;
  defining_relations := empty;
  while there are uncoloured edges do
    choose an uncoloured edge e; add R (e) to defining_relations; colour e;
    repeat (*determine all deductions*)
      for each relation r in defining_relations do
        for each node i do
          trace r around i;
          if loop contains precisely one uncoloured edge f then
            colour f;
          end if;
        end for;
      end for;
    until no more edges can be coloured;
  end while;
end;
```

Example

We will now work through the algorithm for the symmetries of the square, using our previous spanning tree. The uncoloured edges will be omitted, so we can see the effect of the colouring. We begin with the spanning tree, as shown in Figure 5.

Figure 5 : Initially Coloured Edges

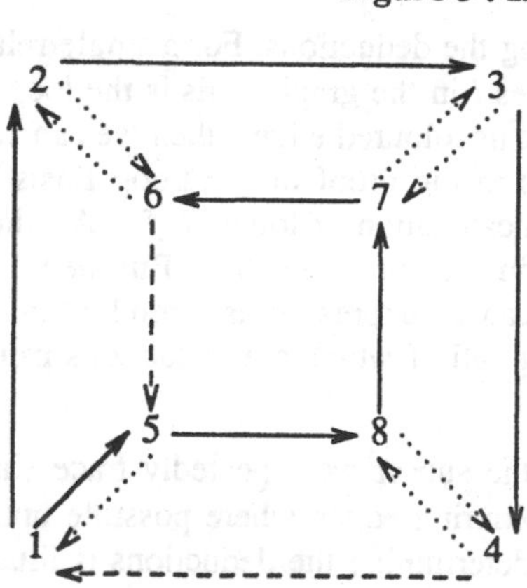

Solid arrow represent coloured edges.

The edge 4 -*a*-> 1 is chosen, and gives the relation a^4. This is the first defining relation. Tracing a^4 around node 5 colours the edge 6 -*a*-> 5. Thus giving Figure 6.

Figure 6 : Coloured Edges after First Defining Relation

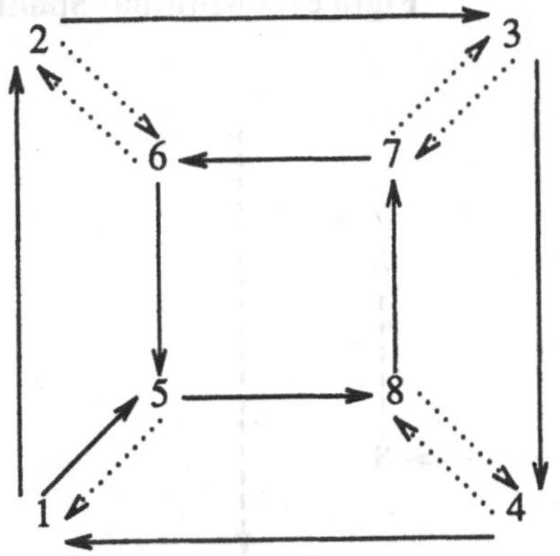

The edge 5 -*b*-> 1 is chosen, and gives the relation b^2. This is added to the set of defining relations. Tracing colours no further edges, and we have Figure 7.

Figure 7 : Coloured Edges after Second Defining Relation

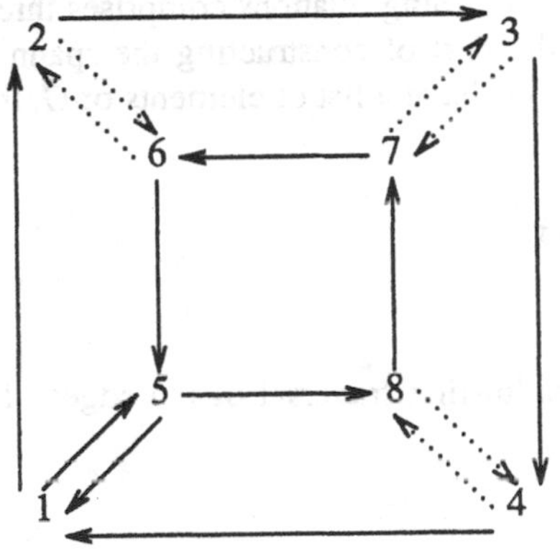

The edge 2 -*b*-> 6 is chosen, and gives the relation $a \times b \times a^{-3} \times b^{-1}$. This is added to the defining relations.
Tracing b^2 around node 2 colours the edge 6 -*b*-> 2.
Tracing $a \times b \times a^{-3} \times b^{-1}$ around node 2 colours the edge 3 -*b*-> 7.
Tracing b^2 around node 3 colours the edge 7 -*b*-> 3.
Tracing $a \times b \times a^{-3} \times b^{-1}$ around node 3 colours the edge 4 -*b*-> 8.
Tracing b^2 around node 4 colours the edge 8 -*b*-> 4.

The graph is completely coloured and the set of defining relations is { a^4, b^2, $a \times b \times a^{-3} \times b^{-1}$ }.

Another Example

In order to obtain short relations we should take a minimal spanning tree such as that shown in Figure 8.

Figure 8 : Minimal Spanning Tree

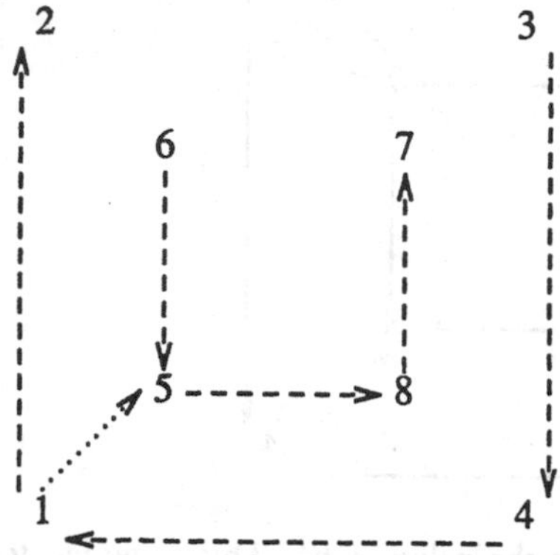

Working through this example we obtain the defining relations { $a^4, b^2, a \times b \times a \times b^{-1}$ }.

Analysis

The total cost of constructing a set of defining relations comprises three components - the cost of constructing the Cayley graph, the cost of constructing the spanning tree, and the cost of the colouring algorithm. Assuming we have a list of elements of G, the cost of constructing a Cayley graph is

$|G| \times |S|$ multiplications, and

$|G| \times |S|$ searches.

Forming the spanning tree is a breadth-first traversal of the edges of the Cayley graph. The cost of this is at most

$|G| \times |S|$ edge traces.

Let L be the sum of the lengths of the defining relations, as constructed by the colouring algorithm. The actual formation of the relations $R(e)$ that make up the set of defining relations requires

L edge traces.

Now let us consider the cost of determining the deductions. Each iteration of the repeat-loop arises from the colouring of at least one edge not in the spanning tree. Hence the number of iterations is at most

$$1 + |G| \times \left[|S| - 1 \right].$$

Each iteration traces the (current) defining relations around all the nodes. This can be no worst than tracing all the defining relations around the nodes, so the total number of edge traces required to determine the deductions is bounded by

$$L \times |G| \times \left[1 + |G| \times \left[|S| - 1 \right] \right].$$

Thus the total worst case cost is proportional to $L \times |G|^2 \times |S|$.

Summary

The Cayley graph is a useful graphical representation of a small group. The relations of the group correspond to loops in the Cayley graph. The colouring algorithm is an effective means of determining deductions from a set of relations, and hence of determining a set of defining relations of the group.

The strategy we have followed for the colouring is the simplest one. It is also the one used in practical implementations. Its worst case cost is dominated by $L \times |G|^2 \times |S|$, where L is the total length of the defining relations.

Exercises

(1/Easy) Not all Cayley graphs are planar, so for the next example we can not nicely draw the graph. Instead we give the abbreviated multiplication table for the group. The entries indicate the product of the row number by the generator at the column heading.

	a	*b*
1	2	7
2	3	6
3	4	5
4	1	8
5	6	1
6	7	4
7	8	3
8	5	2

Thus there are edges 1 -*a*-> 2 and 1 -*b*-> 7 from node 1.

Work through the algorithm for this example. Choose several different spanning trees, if you so desire, and compare the sets of defining relations.

(2/Difficult) The simple colouring strategy often traces relations when there is no possibility for colouring an edge. For example, the loop may already be totally coloured, or, having just traced the loop in the previous iteration and found two uncoloured edges, the strategy would trace it again even if no edges at all have been coloured in the meantime.

The strategy you should analyse follows. For each pair (node, defining relation) keep track of the number of uncoloured edges in the loop. This information is initialised when a defining relation is added to the set by tracing it around all nodes. At other times only trace an edge if there is a possibility of making a deduction and colouring an edge.

Let N be the number of uncoloured edges in the graph. Initially N is $1 + |G| \times \left[|S| - 1 \right]$. If t edges are uncoloured in a loop just traced, then store the integer

0, if $t = 0$, and

$N - t + 1$, if $t \neq 0$.

with the pair (node, defining relation) that corresponds to the loop. The loop is next traced when N is less than or equal to the integer stored with the corresponding pair.

Analyse this strategy.

(3/Moderate) This strategy considers the edges rather than the nodes when tracing relations. A loop will only lead to the colouring of an edge if some other edge in the loop has just been coloured. Keep a list of edges (not in the spanning tree) that have been coloured. Add newly coloured edges to the end of the list.

For each edge in the list, where s is the label of that edge, trace each relation around the edge by matching the edge with each "essentially different" occurrence of s in the relation. For example, the generator a occurs four times in a^4, but each occurrence is the same from the viewpoint of tracing a loop.

If the defining relations are considered to be nearly random, then the average number of essentially different occurrences of a given generator in a set of total length L is $L/|S|$. With this assumption, analyse the strategy to show it is proportional to $L \times |G|$.

Bibliographical Remarks

The use of Cayley graphs to represent groups is widespread in combinatorial group theory. The correspondence between loops and relations is also well known. The classic books H.S.M. Coxeter and W.O.J. Moser, **Generators and Relations for Discrete Groups**, Springer-Verlag, Berlin, 1965, and W. Magnus, A. Karass, and D. Solitar, **Combinatorial Group Theory**, Interscience, New York, 1966 contain much more on Cayley graphs and relations.

The construction of a set of defining relations from a Cayley graph is due to J. J. Cannon, "*Construction of defining relations for finite groups*", Discrete Mathematics **5** (1973) 105-129. This chapter only deals with the single stage algorithm of that paper, and not with the two stage algorithm. While Cannon suggests in his paper that a colouring strategy based on tracing relations around the edges that become coloured may be better than the simplest strategy of tracing relations around nodes, only the simplest strategy has been implemented. Presumably, the simplest strategy is adequate in practice.

It should be noted that the analysis in Cannon's paper is simplified to the point of assuming only a small number of iterations of the **repeat**-loop each time a new defining relation is added to the set.

The paper, John Grover, Lawrence A. Rowe, and Darrell Wilson, "*Applications of coset enumeration*", (Proceedings of the Second Symposium on Symbolic and Algebraic Manipulation, Los Angeles, 1971) (S. R. Petrick, editor), Association of Computing Machinery, New York, 1971, 183-187, describes an interactive approach to constructing defining relations. It constructs a relation $w_1 \times w_2^{-1}$ if the words w_1 and w_2 evaluate to the same element of the group. A coset enumeration is used to check whether the set of relations is defining. A variation of this approach is described in C. M. Campbell and E. F. Robertson, "*Presentations for the simple groups G,* $10^5 < |G| < 10^6$", Communications in Algebra **12**(21) (1984) 2643-2663.

The Schreier-Todd-Coxeter-Sims method briefly mentioned in chapter 13 and described in J. S. Leon, "*On an algorithm for finding a base and strong generating set for a group given by generating permutations*", Mathematics of Computation **35**, 151 (1980) 941-974 determines defining relations for very large permutation groups.

Chapter 6. Lattice of Subgroups

A very complete description of a group is a description of all its subgroups, and their relationship with one another. The subgroups of a group form a lattice. The lattice has layers allowing the subgroups to be generated layer by layer. This approach to their generation is called the cyclic extension method. This chapter presents the ideas behind the cyclic extension method, and then refines the ideas into an effective algorithm. An important space saving device is used to represent subgroups.

Unfortunately, the cyclic extension method cannot construct a certain type of subgroup. These are the perfect subgroups. We discuss briefly how the cyclic extension method could accommodate the perfect subgroups.

Lattices and Layers

A lattice is a set with a partial order, such as the inclusion operator $\leq$ of subgroups, where each pair of members has a least upper bound and a greatest lower bound. For the pair {H, K} of subgroups the least upper bound is the subgroup <H, K>, and the greatest lower bound is the intersection $H \cap K$. The lattice of subgroups of the symmetries of the square is shown in Figure 1.

Figure 1 : Subgroup Lattice of the Symmetries of the Square

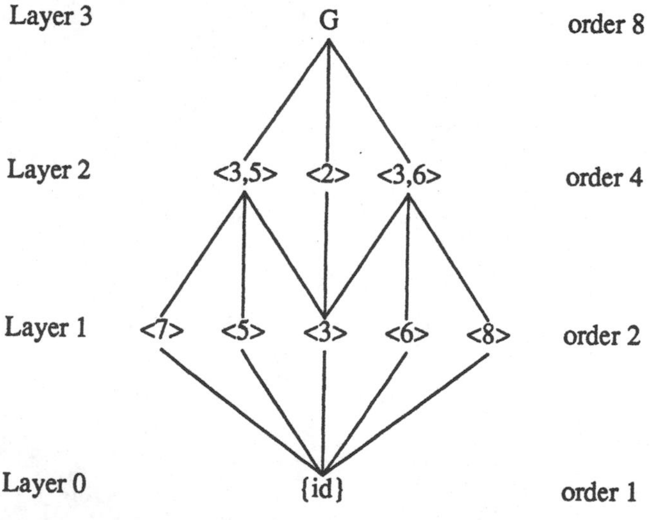

The integers in Figure 1 represent the position of the elements in the original list of elements. The order of the subgroups is noted on the right. Note that the orders (from the top) are the products of three, two, one, and zero primes respectively (even though the primes are not distinct). Hence, we have layers three, two, one, and zero.

The members u of a finite lattice fall naturally into layers, depending on the longest chain of members

$$u_0 < u_1 < \cdots < u_i = u \tag{5.1}$$

that ends at u. Those with chains of length i belong to the i-th layer.

Certain subgroups U of a group have a chain of subgroups

$$id = U_0 < U_1 < \cdots < U_i = U$$

where U_{j-1} is normal in U_j, and the index $|U_j:U_{j-1}|$ is a prime p_j. These subgroups are called *soluble*, and they are amenable to generation by the cyclic extension method. A soluble subgroup in the i-th layer of the subgroup lattice will have an order that is the product of i (not necessarily distinct) primes. The index $|U:U_{i-1}|$ being prime means that the subgroup U can be generated by U_{i-1} and one extra element.

It is convenient to define the layers of the subgroup lattice in terms of the order of the subgroup rather than in terms of the chain (5.1). Thus, a subgroup U belongs to the i-th layer if the order of U is the product of i primes.

Constructing the Next Layer

This section looks at how the cyclic extension method constructs the subgroups in the i-th layer in the lattice from those in the (i-1)-st layer. The main problem is to avoid constructing the same subgroup many times over. First let us characterise the subgroups that do belong to the i-th layer.

A subgroup $U = U_i$ belongs to the i-th layer if there is a subgroup U_{i-1} in the (i-1)-st layer and an element g in the group such that

1. $U_i = < U_{i-1}, g >$,
2. U_{i-1} is normal in U_i, and
3. the index $|U_i:U_{i-1}| = p_i$, a prime.

For this we need an element g not in U_{i-1}, normalizing U_{i-1}, and with g^{p_i} in U_{i-1}, for some prime p_i.

Suppose we have constructed the (i-1)-st layer. To construct the i-th layer, we take each subgroup U_{i-1} in the (i-1)-st layer, find the elements g satisfying the above three conditions, and add $< U_{i-1}, g >$ to the i-th layer if it is not already in the i-th layer.

The obvious algorithm for the inductive step is Algorithm 1.

Algorithm 1 : Obvious Inductive Step

Input : a group G, given by a list of elements;
the (i-1)-st layer of subgroups of G;

Output : the i-th layer of subgroups of G;

```
begin
  for each subgroup U_{i-1} in the (i-1)-st layer do
    for each element g of G do
      if g ∉ U_{i-1} and g^{-1} × U_{i-1} × g = U_{i-1}
        and g^p ∈ U_{i-1}, for some prime p then
        if < U_{i-1}, g > is a new subgroup then
          add < U_{i-1}, g > to i-th layer;
        end if;
      end if;
    end for;
  end for;
end;
```

How to Test a Subgroup is New

For the subgroup $< U_{i-1}, g >$ to be a subgroup W already in the i-th layer, it is necessary and sufficient that $U_{i-1} \subseteq W$ and $g \in W$. If subgroups are represented by bitstrings, then these checks are simple bit operations. Furthermore, they can be performed without generating the subgroup $< U_{i-1}, g >$.

To form the subgroup $U = < U_{i-1}, g >$, we use the fact that U_{i-1} is normal in U and that $g^{p_i} \in U_{i-1}$. Hence, the coset decomposition of U is

$$U = U_{i-1} \cup \left[U_{i-1} \times g \right] \cup \left[U_{i-1} \times g^2 \right] \cup \cdots \cup \left[U_{i-1} \times g^{p_i - 1} \right].$$

So the full power of Dimino's algorithm is not required.

The number of tests to see if subgroups are new can be reduced by never testing elements that do not give a new subgroup. We reorganise the obvious algorithm to use a set Γ which contains the elements of the group that may generate a new subgroup containing U_{i-1}. We know

i. that $U_{i-1} \cap \Gamma$ is empty, and

ii. that $W \cap \Gamma$ is empty, for all subgroups W that contain U_{i-1} and are already in the i-th layer.

These facts lead to Algorithm 2.

Algorithm 2 : Inductive step avoiding old subgroups

```
Input : a group G, given by a list of elements;
        the (i-1)-st layer of subgroups of G;

Output : the i-th layer of subgroups of G;

begin

  for each subgroup U_{i-1} in (i-1)-st layer do

    (*initialise the set Γ to avoid old subgroups*)
    Γ := G - U_{i-1};
    for each subgroup W already in i-th layer do
      if U_{i-1} ⊆ W then Γ := Γ - W; end if;
    end for;

    (*construct new subgroups*)
    while Γ is not empty do

      choose g ∈ Γ;
      if g^{-1} × U_{i-1} × g = U_{i-1} and g^p ∈ U_{i-1}, for some prime p then
        add U = < U_{i-1}, g > to i-th layer;
        Γ := Γ - U;
      else
        Γ := Γ - {g};
      end if;

    end while;

  end for;

end;
```

Note that two of the conditions tested inside the inner loop of Algorithm 1 have been eliminated.

Eliminating the Normality Test

The normality test $g^{-1} \times U_{i-1} \times g = U_{i-1}$ can be eliminated by noting that the elements g which normalize U_{i-1} form a subgroup $N(U_{i-1})$. The search algorithms of Chapter 4 can compute the normalizer more efficiently than stepping through each of the elements of the group - which is basically what the inductive step is doing to find normalizing elements. The modified algorithm is Algorithm 3.

Algorithm 3 : Inductive step using normalizers

```
Input : a group G, given by a list of elements;
    the (i-1)-st layer of subgroups of G;

Output : the i-th layer of subgroups of G;

begin

  for each subgroup U_{i-1} in (i-1)-st layer do

    (*initialise the set Γ to avoid old subgroups*)
    (* and to consider only normalizing elements*)
    compute N(U_{i-1});  Γ := N(U_{i-1}) - U_{i-1};
    for each subgroup W in i-th layer do
      if U_{i-1} ⊆ W then Γ := Γ - W; end if;
    end for;

    (*construct new subgroups*)
    while Γ is not empty do

      choose g ∈ Γ;
      if g^p ∈ U_{i-1}, for some prime p then
        add U = < U_{i-1}, g > to i-th layer;
        Γ := Γ - U;
      else
        Γ := Γ - {g};
      end if;

    end while;

  end for;

end;
```

While it is possible to discard more than just $\{g\}$ from Γ when $g^p \notin U_{i-1}$, for example the powers of g which have the same order as g, we will not follow that up. The next modification will take care of any such improvement.

Using Elements of Prime Power Order

This section shows that only the elements of G whose order is a prime power, p^n, for some prime p and some exponent n, are required as extending elements. The following theorem justifies this claim.

Theorem

Suppose $U = < U_{i-1}, g >$ is in the i-th layer of the subgroup lattice, where U_{i-1} is in the (i-1)-st layer, and

1. U_{i-1} is normal in U, and
2. $g^p \in U_{i-1}$, for some prime p.

Let the order of g be $p^n \times r$, where p and r are coprime. Then g^r has order p^n, has its p-th power in U_{i-1}, and extends U_{i-1} to U.

The proof depends on r and p being coprime. This guarantees the existence of integers a and b such that

$$a \times p + b \times r = 1.$$

The fact that g^r has order p^n follows from $|g| = p^n r$. Since $g^p \in U_{i-1}$ and $(g^r)^p = (g^p)^r$, the p-th power of g^r is in U_{i-1}. The subgroup $< U_{i-1}, g^r >$ is U because g is in this subgroup. Indeed

$$g = g^{ap + br} = (g^p)^a \times (g^r)^b,$$

and $g^p \in U_{i-1}$. Thus completing the proof.

The effect of this result is that only elements g of prime power order are considered when extending the subgroups in layer i-1.

Using Subgroups of Prime Power Order

Amongst the elements of prime power order there is still a lot of redundancy. Suppose g has order p^n. Then the cyclic subgroup $< g >$ consists of the subgroup $< g^p >$ and the powers of g that have order p^n. The latter elements are g^r, where r and p are coprime. Since $<g> = <g^r>$, we have $<U_{i-1}, g> = <U_{i-1}, g^r>$, and so g will extend U_{i-1} to a subgroup U in the i-th layer if and only if g^r extends U_{i-1} to exactly the same subgroup U. Therefore it suffices to consider precisely one generator from each cyclic subgroup of prime power order.

Let Z be the set of cyclic subgroups of G of prime power order. For the symmetries of the square,

$$Z = \{ <2>, <3>, <5>, <6>, <7>, <8> \}.$$

Let *Z_generators* be a set of generators of the cyclic subgroups in Z. That is, choose one generator for each subgroup. For the symmetries of the square we have,

$$Z\_generators = \{ elt[2], elt[3], elt[5], elt[6], elt[7], elt[8] \}.$$

Typically *Z_generators* is much smaller than the whole group G.

The inductive step of the lattice algorithm only needs to scan the elements of *Z_generators* rather than G for elements satisfying the conditions. The algorithm is Algorithm 4.

Algorithm 4 : Inductive step using cyclic subgroup generators

Input : a group G, given by a list of elements;
the (i-1)-st layer of subgroups of G;
a set *Z_generators* of generators of the cyclic subgroups of G of prime power order;

Output : the i-th layer of subgroups of G;

```
begin
  for each subgroup U_{i-1} in (i-1)-st layer do

    (*initialise the set Γ to avoid old subgroups*)
    (* and to consider only normalizing elements*)
    (* and to use only the representative elements of prime power order *)
    compute N(U_{i-1}); Γ := Z_generators ∩ {N(U_{i-1}) - U_{i-1}};
    for each subgroup W in i-th layer do
      if U_{i-1} ⊆ W then Γ := Γ - W; end if;
    end for;

    (*construct new subgroups*)
    while Γ is not empty do
      choose g ∈ Γ;
      if g^p ∈ U_{i-1}, for some prime p then
        add U = < U_{i-1}, g > to i-th layer; Γ := Γ - U;
      else
        Γ := Γ - {g}; (*this is now the best possible*)
      end if;
    end while;

  end for;

end;
```

There is only one prime relevant to each element in *Z_generators*, so the corresponding power g^p would already be known. Thus the remaining test in the inner loop is a simple bit operation.

The much smaller size of *Z_generators* than G makes our previous decision to compute normalizers by the search algorithm, rather than test that the candidates normalize U_{i-1}, suspect. The number of candidates is possibly quite small for any given subgroup U_{i-1}. Perhaps it is cheaper to test if they normalize U_{i-1} than to generate the whole normalizer. In practice, the normalizer is computed, but this decision is influenced by the fact that the normalizer of each subgroup is considered part of the description of the subgroup lattice. Thus it would be computed even if it were not used in the inductive step.

Full Lattice Algorithm for Soluble Subgroups

The first task of the full algorithm for computing the lattice of soluble subgroups is to calculate the cyclic subgroups of prime power order, and to choose a set of generators. This can be done by running through the elements of the group. The first layer will consist of the subgroups of order p, for some prime p. The cyclic subgroups of order p^n can be the first subgroups included in the n-th layer, if one so desires. If the order of the group is $\prod_{i=1}^{r} p_i^{n_i}$, then the last layer constructed is layer $(\sum_{i=1}^{r} n_i) - 1$. There is no need to construct the very top layer, since it consists of only G.

Analysis

This section analyses the full algorithm using each of the algorithms for the inductive step. Our major assumption is that bit operations are so cheap relative to element multiplications that they are performed at zero cost.

The cost of testing if an element g normalizes the subgroup U_{i-1} is $2 \times (i-1)$ multiplications to form the conjugates of the generators of U_{i-1}, and $(i-1)$ searches to find the position of the conjugates in the list of elements of G. This is equivalent to a total of $4 \times (i-1)$ multiplications.

The cost of testing $g^p \in U_{i-1}$ for all primes p involved in the order of g is small. We will approximate it by the constant c (about 8) multiplications per test.

The cost of forming a subgroup U as a bitstring is equivalent to $3 \times |U|$ multiplications, since we know the coset representatives of U_{i-1}. Although the subgroups involved in a layer will vary in size, this distinction will not be made in the formulae. Think of U_i as representing the "average" subgroup in layer i.

Let L_i be the number of subgroups in the i-th layer, and suppose that G belongs to the *top*-th layer. The total cost when using Algorithm 1 as the inductive step, and when all subgroups $< U_{i-1}, g >$ are formed is bounded by

$$\sum_{i=0}^{top-2} \left[L_i \times |G| \times \left[3 \times |U| + 4 \times i + c \right] \right]$$

multiplications because we extend layers 0 to *top*-2.

The total cost when using Algorithm 2 is bounded by

$$3 \times \sum_{U<G} |U| + \sum_{i=0}^{top-2} \left[L_i \times |G| \times \left[4 \times i + c \right] \right]$$

multiplications because now each subgroup is formed precisely once.

The total cost when using Algorithm 3 is bounded by

$$3 \times \sum_{U<G} |U| + \sum_{i=0}^{top-2} \left[L_i \times \left[cost\ of\ N(U_{i-1}) + c \times |N(U_{i-1})| \right] \right]$$

multiplications.

The cost of forming Z by determining the powers of each element is one multiplication and one comparison per element, because we deal with the powers of an element at the same time we deal with the element itself. The total cost when using algorithm 4 is bounded by

$$2 \times |G| + 3 \times \sum_{U<G} |U| + \sum_{i=0}^{top-2} \left[L_i \times cost\ of\ N(U_{i-1}) \right]$$

multiplications. If we run through *Z_generators* testing the normalizing condition rather than construct the normalizer then the cost is bounded by

$$2 \times |G| + 3 \times \sum_{U<G} |U| + 4 \times |Z| \times \sum_{i=0}^{top-2} \left[L_i \times i \right]$$

multiplications.

Saving Space

A group may have very many subgroups. So many in fact, that representing them as a characteristic function of the list of elements of the group will require a large amount of space. While such a characteristic function is needed to represent an arbitrary *subset*, an arbitrary *subgroup* can be represented more compactly as a characteristic function of the set Z of cyclic subgroups of prime power order. That is, a subgroup U is uniquely determined by the cyclic subgroups of prime power order that it contains. In fact, the elements of the subgroup are just the products of those in the cyclic subgroups of prime power order. If we list the subgroups of Z, or alternatively, list the elements of *Z_generators*, then a subgroup can be represented as a characteristic function of this list.

This still allows us to perform efficient bit operations to test inclusion of subgroups, form differences, and only slightly complicates forming $< U_{i-1}, g >$, since we must convert from the list of elements in the subgroup to the cyclic subgroups in the subgroup. This involves determining which elements of *Z_generators* are in the subgroup - these are simple bit operations.

In general, the size of Z is much smaller than the size of G, so this is a considerable saving in space.

A cyclic subgroup of prime power order is referred to as a *zuppo*, an acronym from the original German work on subgroup lattices.

Perfect Subgroups

The subgroups which cannot be constructed by the cyclic extension method are those with a chain

$$id \neq U_0 < U_1 < \cdots < U_i = U$$

(where U_{i-1} is normal of prime index in U_i) that cannot be prolonged as far down as the identity subgroup. The subgroup U_0 in these cases is a *perfect* subgroup. Once the perfect subgroups are included in the lattice, then the cyclic extension method will construct the other subgroups in the chain. So the problem is detecting and including the perfect subgroups.

All perfect subgroups of a group are contained in the unique largest perfect subgroup of the group. This is the last term in what is called the *derived series* of the group. The derived series can be constructed without difficulty, though we shall not discuss it here. There is a list of all perfect groups up to order 10 000 together with information on their structure, and the perfect subgroups they contain. The information is sufficient to easily distinguish any two perfect groups, and to reconcile the notation of the list with the actual description of the perfect group as a subgroup of G. Using this information, we can locate the last term of the derived series in the list, and include all its perfect subgroups into the lattice. The order of a perfect subgroup indicates where it should be added to the lattice of soluble subgroups.

Summary

With a knowledge of all the perfect groups, the cyclic extension method can determine all the subgroups of a group, and arrange them as a lattice. Clever use of bit operations and reducing the extending elements considered from the whole group to just the generators of the cyclic subgroups of prime power order saves a lot of time, and the use of characteristic functions of zuppos saves a lot of space.

Exercises

(1/Moderate) For the group in exercise 1 of Chapter 5, construct the subgroup lattice. It has no perfect subgroups.

(2/Moderate) For the symmetric group of degree 4 construct the subgroup lattice. It has no perfect subgroups.

(3/Moderate) Consider the following means of determining all the subgroups. Enumerate the $2^{|Z|}$ subsets S of *Z_generators*. For each subset S determine $<S>$ (as a characteristic function of the zuppos) using Dimino's algorithm. If $<S>$ is not already in the list of subgroups then add it to the list.

Execute this algorithm on the symmetries of the square.

(4/Moderate) Modify the algorithm of exercise 3 by keeping a set F of "filters". That is, if $<S> = G$ then add S to F. Before constructing $<S>$ test if there is some filter S' contained in S. If there is such a filter then we know that $<S> = G$.

Execute this algorithm on the symmetries of the square.

(5/Moderate) Order the enumeration of subsets S of *Z_generators* so that the subsets T containing S come in a consecutive sequence after S. Use this order of enumeration to improve the algorithm of exercise 4. That is, if we find a filter S' contained in S then the filter is also contained in all subsets containing S. Hence, if the filters eliminate S we should automatically eliminate all subsets T containing S.

Execute this algorithm on the symmetries of the square.

(6/Moderate) The order of enumeration in exercise 5 allows us to improve the computation of $<S>$. Suppose $S = \{s_1, s_2, \ldots, s_m\}$. Then $S' = \{s_1, s_2, ..., s_{m-1}\}$ comes before S in the order of enumeration. If we have not eliminated S using the filters then S' was also not eliminated. Hence, $<S'>$ is some subgroup in the list of subgroups. Organize the algorithm to keep track of its location, and use the subgroup in constructing $<S>$.

The subgroup $<S'>$ can be used in two ways.

(i) if $s_m \in <S'>$ then there is no need to construct $<S>$ since this subgroup is $<S'>$.

(ii) if $s_m \notin <S'>$ then use the elements of $<S'>$ in the inductive step of Dimino's algorithm to compute $<S>$.

Modify the previous algorithm and execute it on the symmetries of the square.

(7/Moderate) A *maximal* subgroup H of G is a subgroup of G which has no larger subgroup U between H and G. That is, there is no subgroup U such that $H < U < G$.

Suppose that H is maximal and that $<S'> = H$. Then $S' \subseteq S$ implies that $<S>$ is H or G.

The set of filters F can be generalized to include subsets S' that generate G or a maximal subgroup (and this maximal subgroup has already been added to the list of subgroups).

How do we tell if $<S'>$ is maximal? In general we cannot (before we have constructed all the subgroups). However, there are some important special cases. If the index of H in G is a prime then H is maximal. So if we construct $<S'> = H$ with prime index then we can add S' to the set of filters.

Modify the algorithm of exercise 5 to incorporate this change and execute it on the symmetries of the square.

Bibliographical Remarks

The computation of subgroup lattices began in 1958 with the development and implementation of the cyclic extension method by Joachim Neubüser in Kiel. His work is published in J. Neubüser, "*Untersuchungen des Untergruppenverbandes endlicher Gruppen auf einer programmgesteuerten elektronischen Dualmaschine*", Numerische Mathematik **2** (1960) 280-292. The work has been continued by Neubüser and his students. The references are V. Felsch, **Ein Programm zur Berechnung der Untergruppenverbandes und der Automorphismengruppe einer endlichen Gruppe**, Diplomarbeit, Kiel, 1963; V. Felsch and J. Neubüser, *Ein Programm zur Berechnung des Untergruppenverbandes einer endlichen Gruppe*, Mitteilungen des Rheinisch-Westfälischen Institutes für Instrumentelle Mathematik (Bonn) **2** (1963) 39-74; P. Dreyer, **Ein Programm zur Berechnung der auflösbaren Untergruppen von Permutationsgruppen**, Diplomarbeit, Kiel, 1970; H. Steinmann, **Ein transportables Programm zur Bestimmung des Untergruppenverbandes von endlichen auflösbaren Gruppen**, Diplomarbeit, Aachen, 1974; Bernd Bohmann, **Ein Verfahren zur Bestimmung von Untergruppenverbänden, das die Benutzung von Sekundärspeicher zulässt, und seine Implementation**, Diplomarbeit, Aachen, 1978.

The program of Felsch (in 1963) contained information about the five smallest perfect groups. This was sufficient to complete the subgroup lattice of any group within the program's storage capability. The other programs mentioned above only determined the soluble subgroups. The list of perfect groups (up to order 10 000) was determined in G. Sandlöbes, "*Perfect groups of order less then* 10^4", Communications in Algebra, **9** (1981) 477-490 and used in the implementation described in V. Felsch and G. Sandlöbes, "*An interactive program for computing subgroups*", **Computational Group Theory**, (Proceedings of LMS Symposium on Computational Group Theory, Durham, 1982), edited by M. Atkinson, Academic Press, 1984, 137-143.

The construction of subgroups in layers really requires very few subgroups to actually be in memory. Indeed only information about the subset Γ, and the subgroups U and U_{i-1}, as well as information about the elements of G and the cyclic subgroups in Z are required. The subgroups of the (i-1)-st and i-th layer are processed sequentially, so their storage on disc is straightforward. In Neubüser's original implementation memory was synonymous with magnetic drums. Felsch in 1965 extended his own implementation to optionally use two magnetic tapes as external storage. Space was still restrictive as all the subgroups of a layer were required to fit into memory. While both Dreyer and Steinmann allowed the use of external storage, it was Bohmann who reorganised the cyclic extension method to no longer build the lattice layer by layer and thereby properly use external storage.

In practice, the lattice is augmented by information about normalizers and centralizers of subgroups, the conjugacy of subgroups, and other properties of subgroups. This represents a wealth of information about the group. Recent work has dealt with the problem of tailoring and interrogating this information, as it is too easy to be swamped by such a mass of detail. This work is described in the paper of V. Felsch and G. Sandlöbes.

There is an alternative to the cyclic extension method that was developed and implemented by L. Gerhards and W. Lindenberg, "*Ein Verfahren zur Berechnung des vollständigen Untergruppenverbandes endlicher Gruppen auf Dualmaschinen*", Numerische Mathematik **7** (1965) 1-10; and W. Lindenberg and L. Gerhards, "*Combinatorial construction by computer of a set of all subgroups of a finite group by composition of partial sets of its subgroups*", **Computational Problems in Abstract Algebra** (Proceedings of a conference, Oxford, 29 August - 2 September, 1967), J. Leech (editor), Pergamon Press, Oxford, 1970, 75-82. The main ideas of the method are presented in exercises 3-7. It is a combinatorial approach that has the advantage of constructing all the subgroups, including the perfect ones. As with the cyclic extension method, there is no adequate analysis of this method. Empirical timings indicate the method is competitive with the cyclic extension method for small groups with orders less than a few hundred. However, when the order of the group exceeds one thousand it suffers "combinatorial explosion" and is very slow. The cyclic extension method copes with groups with orders in the tens of thousands.

The determination of the derived series is described in J.J. Cannon, "*Computing local structure of large finite groups*", **Computers in Algebra and Number Theory** (Proceedings of a Symposium on Applied Mathematics, New York, 1970), G. Birkhoff and M. Hall, Jr (editors), SIAM-AMS Proceedings, volume **4**, American Mathematical Society, Providence, R.I., 1971, pp 161-176; and G. Butler and J.J. Cannon, "*Computing in permutation and matrix groups I: normal closure, commutator subgroups, series*", Mathematics of Computation, **39**,160 (Oct. 1982) 663-670.

Chapter 7. Orbits and Schreier Vectors

This chapter defines the concepts of orbit and Schreier vector and presents algorithms for computing them. The main question concerning the permutation group that they allow us to answer is: Given two points δ and γ, is there an element mapping δ to γ, and , if so, determine such an element.

Orbits

Let G be a permutation group acting on the set $\Omega = \{1, 2, ..., n\}$ of points. Let G be generated by the set $S = \{s_1, s_2, ..., s_m\}$.

The orbits of G on Ω indicate how G acts on Ω by telling us whether there is an element mapping one point to another. There are three (equivalent) views of orbits. We will present all three.

The *orbit* of G containing the point δ is the set

$$\delta^G = \{ \delta^g \mid g \in G \}$$

of images of δ under elements of the group G.

Define the relation $\tilde{}$ on the set of points by : $\delta \tilde{} \gamma$ if and only if there is an element g of G such that $\delta^g = \gamma$.

This defines an equivalence relation since it is

i. reflexive (for all $\delta \in \Omega$, $\delta^{id} = \delta$),

ii. symmetrical (if $\delta^g = \lambda$ *then* $\lambda^{g^{-1}} = \delta$), and

iii. transitive (if $\delta^g = \lambda$ *and* $\lambda^h = \alpha$, then $\delta^{g \times h} = \alpha$).

The equivalence classes partition the set Ω of points. An *orbit* is an equivalence class of the relation $\tilde{}$.

The generators S determine a (labelled, directed) graph (possibly with loops) as follows. The vertices are the points of Ω. There is an edge

$$\delta \xrightarrow{\quad s \quad} \gamma$$

in the graph if a generator s maps δ to γ. An *orbit* of G on Ω is the set of vertices of a connected component of the graph. (This definition is independent of the generating set of G.)

It is this last view of an orbit that suggests an algorithm for computing an orbit. Algorithm 1 that determines the orbit δ^G is a breadth-first traversal of the connected component containing δ. This is the same as closing the set $\{ \delta \}$ under the operation of taking images.

Algorithm 1 : Determining one orbit

Input : a set $S = \{s_1, s_2, ..., s_m\}$ of generators of a group G acting on Ω;
    a point δ;

Output : the orbit δ^G;
begin
  orbit := { δ };
  for each point γ in *orbit* **do**
    for each generator s in S **do**
      if γ^s not in *orbit* **then** add γ^s to *orbit*; **end if**;
    end for;
  end for;
end;

As an example, consider the group G acting on $\Omega = \{1,2,...,11\}$ that is generated by

$$a = (1,4,5,11,6,10,3,2)(7,8)$$
$$b = (1,4,5,11,6,10,3,2)(8,9)$$
$$c = (1,4,2,3)(5,11,10,6)$$

The graph we obtain is shown in Figure 1.

Figure 1 : Graph Depicting Action of Generators

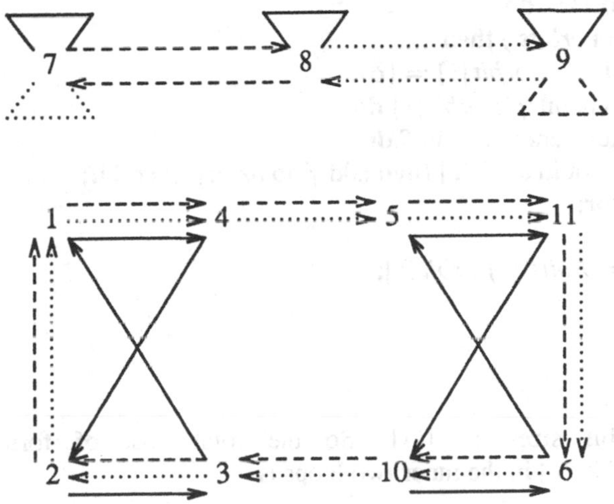

The generator a is represented by dashed edges.
The generator b is represented by dotted edges.
The generator c is represented by solid edges.

Algorithm 1 calculates the orbit of $\delta = 7$ as follows: Initially *orbit* = {7}. With γ=7, γ^a=8, so 8 is added to *orbit* giving {7,8}; γ^b=7; γ^c=7. With γ=8, γ^a=7; γ^b=9, so 9 is added to *orbit* giving {7,8,9}; γ^c=8. With γ=9, γ^a=9; γ^b=8, γ^c=9. Hence, the result is {7,8,9}.

Calculating the orbit of $\delta = 1$ adds the points to the orbit in the sequence 1,4,5,2,11,3,6,10.

The analysis of the algorithm shows that

- $|orbit| \times |S|$ images γ^s are formed,
- $|orbit| \times |S|$ tests of membership in the orbit are performed, and
- $|orbit|$ points are added to the orbit.

If the orbit is represented as a characteristic function on the set Ω, then both the membership test and the addition of a point to the orbit require one operation. The cost of initializing the orbit is $|\Omega|$ operations. If permutations are represented in image form then the calculation of an image requires one operation. Under these assumptions the total cost is

$$|orbit| \times \left[2 \times |S| + 1\right] + |\Omega| \text{ operations.}$$

Algorithm 2, which computes all the orbits of a group, is a simple extension of Algorithm 1.

Algorithm 2 : All orbits

```
Input : a set S = {s_1, s_2, ..., s_m} of generators for a group G acting on Ω;
Output : the orbits orbit[1], orbit[2],... of G on Ω;
begin
  orbits := empty;  i := 0;
  for δ := 1 to |Ω| do
    if not ( δ in orbits ) then
      i := i + 1;       orbit[i] := {δ};
      for each point γ in orbit[i] do
        for each generator s in S do
          if γ^s not in orbit[i] then add γ^s to orbit[i]; end if;
        end for;
      end for;
      orbits := orbits ∪ orbit[i];
    end if;
  end for;
end;
```

The sum of the orbit sizes is $|\Omega|$. So the total cost of forming the orbits *orbit* [1], *orbit* [2],...,*orbit* [N] in the inner two loops is

$$|\Omega| \times \left[2 \times |S| + 1\right] + N \times |\Omega| \text{ operations,}$$

where N is the number of orbits. To this we must add the cost of the tests on δ, the cost of initializing *orbits*, and the cost of forming the union of the orbits in *orbits*. If points are added to *orbits* as they are added to *orbit* [i] then the total cost of the these tasks is $3 \times |\Omega|$ operations. Hence, the total cost of Algorithm 2 is

$$|\Omega| \times \left[2 \times |S| + N + 4\right] \text{ operations.}$$

Schreier Vectors

While the orbits tell us whether there is an element mapping one point to another, it is the Schreier vectors that determine such an element.

Consider the graph constructed from a set of generators for a permutation group acting on Ω. We may consider the edges of the graph to be bi-directional because the edge

$$\alpha \longleftarrow s \longrightarrow \beta$$

is equivalent to the edge

$$\alpha \longrightarrow s^{-1} \longrightarrow \beta$$

Hence, we may assume the edge is in the desired direction. If the points δ and γ are in the same connected component then there is a path

$$\delta = \alpha_0 \text{ --- } s_{i_1} \longrightarrow \alpha_1 \text{ --- } s_{i_2} \longrightarrow \alpha_2 \text{ --- } s_{i_3} \longrightarrow \cdots \text{ --- } s_{i_r} \longrightarrow \alpha_r = \gamma$$

from δ to γ. Then the element

$$s_{i_1} \times s_{i_2} \times \cdots \times s_{i_r}$$

is in the group G and maps δ to γ.

A spanning forest of the graph provides a unique path from the root of a tree to any point in the connected component of the root. A spanning forest of the previous example is shown in Figure 2.

Figure 2 : Spanning Forest

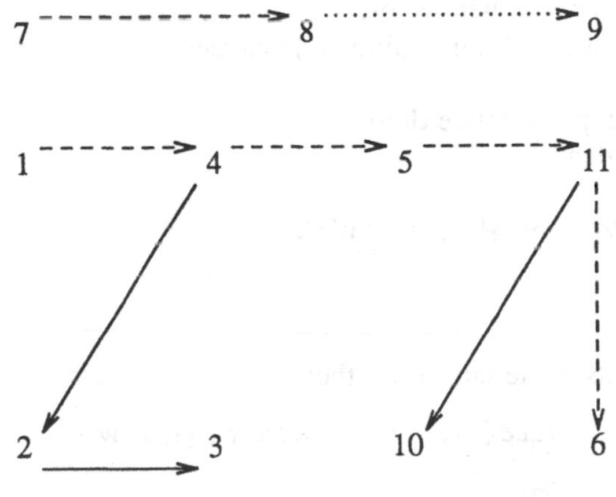

The generator a is represented by dashed edges.
The generator b is represented by dotted edges.
The generator c is represented by solid edges.

A *Schreier vector* of an orbit (or orbits) relative to a set S of generators is a means of representing a spanning tree (forest) of the connected component(s) that corresponds to the orbit(s). The main information we need is the label of the edge leading to a point γ. A Schreier vector stores the edge labels in a vector indexed by the points of Ω. The entry for a point γ is the generator (or its inverse) that labels the edge leading into γ. For example,

	1	2	3	4	5	6	7	8	9	10	11
v		c	c	a	a	a		a	b	c	a

where the orbit representatives (corresponding to the roots of the trees) have no entry. The father of a point γ in the spanning tree is the image of γ under the inverse of $v[\gamma]$. These are often stored in another vector, called the ***backward pointers***. For example,

	1	2	3	4	5	6	7	8	9	10	11
w		4	2	1	4	11		7	8	11	5

To determine an element mapping the orbit representative to a point γ in the connected component, we use Algorithm 3.

Algorithm 3 : Tracing a Schreier vector

Input : a Schreier vector v and backward pointers w of the orbits of a group G;
a point γ;

Output : an element mapping the orbit representative to γ;

```
function trace(γ : point;
               v : Schreier vector;
               w : backward pointers ) : element;
begin
  if γ is orbit representative then
    result is identity;
  else
    result is trace( w[γ], v, w ) × v[γ];
  end if;
end;
```

If γ_1 and γ_2 are two points in the same orbit then

$$trace(\gamma_1, v, w)^{-1} \times trace(\gamma_2, v, w)$$

is an element mapping γ_1 to γ_2.

If the backward pointers are not known explicitly then they can be calculated from the relationship

$$w[\gamma] = \gamma^{v[\gamma]^{-1}}.$$

Of course, this requires us to know (or calculate) the inverses of the generators. We will often abuse the function *trace* by using only the first one or two arguments. In these cases we will be unconcerned whether the backward pointers are known explicitly, or we will assume the the existence of the Schreier vector.

The Schreier vector (and backward pointers) of *one* orbit can be determined by suitably modifying Algorithm 1. We present the modified algorithm as Algorithm 4.

Algorithm 4 : Determining an orbit and Schreier vector

```
Input : a set S = {s_1, s_2, ..., s_m} of generators of a group G acting on Ω;
        a point δ;

Output : the orbit δ^G;
         a Schreier vector v and backward pointers w of the orbit δ^G
                relative to the set S of generators;
begin
  orbit := { δ };
  for γ := 1 to |Ω| do v[γ] := 0;    w[γ] := 0; end for;

  for each point γ in orbit do
    for each generator s in S do
      if γ^s not in orbit then
        add γ^s to orbit;  v[γ^s] := s;  w[γ^s] := γ;
      end if;
    end for;
  end for;
end;
```

Similarly to the analysis of Algorithm 1, we see that the total cost of Algorithm 4 is

$$|orbit| \times \left[2 \times |S| + 3 \right] + 3 \times |\Omega| \text{ operations.}$$

Executing Algorithm 4 with our previous group $G = <a, b, c>$ on $\Omega = \{1,2,...,11\}$ with $\delta = 1$ gives

	1	2	3	4	5	6	7	8	9	10	11
v		c	c	a	a	a				c	a
w		4	2	1	4	11				11	5

This corresponds to the spanning tree illustrated earlier.

Note that, by virtue of the breadth-first traversal, Algorithm 4 constructs the Schreier vector corresponding to a *minimal* spanning tree of the connected component.

All the orbits and a Schreier vector for a spanning forest can be determined by suitably modifying Algorithm 2.

Analysis of Algorithm 3

The cost of determining an element mapping γ_1 to γ_2 is dependent upon the depth of the spanning tree. Let $d(\gamma)$ be the depth of γ in the spanning tree, then the call *trace* (γ, v, w) requires $d(\gamma)$ multiplications and $2 \times d(\gamma)$ operations. Each multiplication requires $2 \times |\Omega|$ operations. Thus the total cost is

$$2 \times d(\gamma) \times \left[\, |\Omega| + 1 \, \right] \text{ operations.}$$

The cost of determining an element mapping γ_1 to γ_2 is

$$2 \times \left[d(\gamma_1) + d(\gamma_2) \right] \times \left[\, |\Omega| + 1 \, \right] \text{ operations.}$$

The greatest value for the depth of a point in a spanning tree is the size of the orbit. This occurs for the cyclic group $G = \langle s \rangle$, where the spanning tree is linear, as in Figure 3.

Figure 3 : Spanning Tree of Cyclic Group

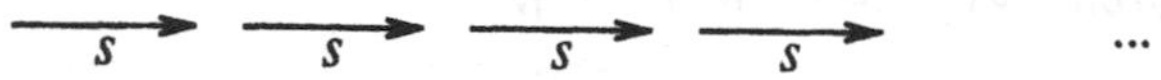

The *best* worst case we can expect is a spanning tree of depth $\log_{|S|}(\, |orbit| \,)$. In general, any tree can be a spanning tree. More accurately,

Proposition

Let T be a tree with n nodes, where each node has at most m children. Then there are permutations $s_1, s_2, \ldots, s_m$ of $\Omega = \{1, 2, ..., n\}$ and a labelling of the nodes of T and the edges of T such that Algorithm 4 using $S = \{s_1, s_2, ..., s_m\}$ as generators and $\delta = 1$ produces the labelled tree T as the spanning tree.

Alas, we have no idea how many choices of the permutations and labellings there are for a given tree. So we do not know the distribution of spanning trees, and hence do not know the average depth of a point in the spanning tree.

Summary

Orbits and Schreier vectors allow us to decide whether there is an element mapping one point to another, and to determine such an element, if it exists. The cost of determining all the orbits and Schreier vectors is $O(\, |\Omega| \times |S| \,)$. The cost of determining an element mapping one point to another is $O(\, |\Omega|^2 \,)$.

Exercises

(1/Easy) For the groups given in chapter 2, calculate their orbits and Schreier vectors.

(2/Easy) Using the Schreier vector calculated in the example of Algorithm 4, determine an element mapping the point 2 to the point 11.

3/Moderate) A group G acts on its elements by conjugation. The orbits in this case are called *conjugacy classes*. The elements in a conjugacy class have the same order. This information is useful in finding the zuppos - the initial step of the subgroup lattice algorithm. Use Algorithm 2 to determine the conjugacy classes of the symmetric group of degree 4. Take the generators to be $\{(1,2,3,4), (1,2)\}$.

Bibliographical Remarks

The algorithms of this chapter are adaptations of graph theoretic algorithms for determining closure and spanning trees.

The concept of an orbit is a very old one. The computational significance of them was recognized by Charles Sims, who also introduced the concept of a Schreier vector. The algorithms to compute them and to trace them occur in lectures that Sims gave in Oxford in January and February 1973. The algorithms appear in J. S. Leon, "*On an algorithm for finding a base and strong generating set for a group given by generating permutations*", Mathematics of Computation **35**, 151 (1980) 941-974.

Orbits can also be computed as partition joins. One partition join algorithm for computing orbits is presented in J. S. Leon, "*An algorithm for computing the automorphism group of a Hadamard matrix*", Journal of Combinatorial Theory, series A **27**, 3 (1979) 289-306. Asymptotically these algorithms are $O(|\Omega| \times log(|\Omega|) + |\Omega| \times |S|)$. One can also adapt union-find algorithms to compute orbits and achieve a complexity which is essentially linear in $|\Omega|$ and $|\Omega| \times |S|$.

The view of a Schreier vector as a tree is first presented in C. M. Hoffman, **Group-theoretic Algorithms and Graph Isomorphism**, Springer-Verlag, Berlin, 1982, which also presents algorithms for constructing orbits and Schreier vectors.

Chapter 8. Regularity

This chapter is concerned with the question: Is there a permutation (other than the identity) in the group which fixes some point? We wish to answer this question using just the generators of the group. We do not want to look at the individual elements of the group.

The algorithms of this chapter illustrate the concepts of orbits and Schreier vectors. In particular, the use made of Schreier vectors is a typical one. It is hoped that this chapter will reinforce the reader's understanding of these important concepts.

Definitions and a First Algorithm

A permutation group $G = \langle s_1, s_2, \cdots s_m \rangle$ is *regular* if Ω is an orbit of G and there is no element of G (other than the identity) that fixes a point.

For example, the group of degree 8 generated by a=(1,2,3,4)(5,6,7,8) and b=(1,5)(2,8)(3,7)(4,6) is regular. It is not difficult to see that Ω is an orbit. The group has precisely eight elements,

identity
(1,2,3,4)(5,6,7,8)
(1,3)(2,4)(5,7)(6,8)
(1,4,3,2)(5,8,7,6)
(1,5)(2,8)(3,7)(4,6)
(1,8)(2,7)(3,6)(4,5)
(1,7)(2,6)(3,5)(4,8)
(1,6)(2,5)(3,8)(4,7)

none of which (other than the identity) fixes any point.

A permutation group is *semiregular* if there is no element (other than the identity) that fixes a point.

The group of degree 8 generated by a=(1,2)(3,4)(5,6)(7,8) and b=(1,3)(2,4)(5,7)(6,8) is semiregular. The group has precisely four elements

identity
a = (1,2)(3,4)(5,6)(7,8)
b = (1,3)(2,4)(5,7)(6,8)
$a \times b$ = (1,4)(2,3)(5,8)(6,7)

none of which (other than the identity) fixes any point. The group is not regular because its orbits are {1,2,3,4} and {5,6,7,8}.

The theoretical fact on which we base our algorithm is

Lemma

Let α be any point in Ω. If Ω is an orbit of G, and if, for each generator s of G there is a permutation z of Ω such that $\alpha^z = \alpha^s$ and such that z commutes with all the generators of G, then G is regular.

Suppose α is the representative of the orbit Ω of G, and that v is a Schreier vector of the orbit with respect to the generators S. Since the permutation z of the lemma should commute with each generator, it should therefore commute with each element of the group. In particular, z should commute with $trace(\beta, v)$, for all points β. Hence,

$$\beta^z = (\alpha^{trace(\beta,\, v)})^z = \alpha^{trace(\beta,\, v) \times z} = \alpha^{z \times trace(\beta,\, v)} = (\alpha^z)^{trace(\beta,\, v)}.$$

However, the other property of z is that we know α^z. Hence, the above equation determines the image of every point under the action of z. It remains to verify that the action defines a permutation, and that the permutation commutes with all the generators of G. The algorithm is presented as Algorithm 1.

Algorithm 1 : Test regularity

Input : a set S of generators of a permutation group G on Ω;

Output : whether G is regular or not;

begin

  form the orbits and Schreier vector v of G;
  if Ω is not an orbit **then exit** with result "not regular"; **end if**;

  α := orbit representative;
  for each generator s of G **do** (* form z such that $\alpha^z = \alpha^s$ *)
    $\gamma := \alpha^s$;
    for each point β in Ω **do**
      $image := \gamma^{trace(\beta,\, v)}$; $z[\beta] := image$
    end for;
    if (z is not a permutation)
        or (z does not commute with all generators) **then**
      exit with result "not regular";
    end if;
  end for;

  exit with result "regular";
end;

We may as well check that z is a permutation as we construct it. To do this we check whether each point is used as an image precisely once. The modified algorithm is Algorithm 2.

Algorithm 2 : Test regularity

Input : a set S of generators of a permutation group G on Ω;

Output : whether G is regular or not;

```
begin

  form the orbits and Schreier vector v of G;
  if Ω is not an orbit then exit with result "not regular"; end if;

  α := orbit representative;
  for each generator s of G do (* form z such that α^z = α^s *)

    γ := α^s;
    for each point β in Ω do used[β] := false; end for;
    for each point β in Ω do
      image := γ^trace(β, v);
      if used[image] then
        exit with result "not regular";
      else
        z[β] := image;  used[image] := true;
      end if;
    end for;

    if z does not commute with all generators then
      exit with result "not regular";
    end if;

  end for;

  exit with result "regular";
end;
```

The main components of the cost of Algorithm 2 are the $|\Omega| \times |S|$ calls to *trace*, and checking commutativity. The call to *trace* does not need to return an element of the group. All we are interested in is the action of the element on the point γ. Thus *trace* can be modified, and the worst case cost of the modified algorithm is $3 \times |\Omega|$. To check if two permutations commute, we must check for each point that the image under their composition is the same irrespective of the order of composition. This requires $4 \times |\Omega|$ operations. Thus to check the commutativity of each z with each generators costs $4 \times |S|^2 \times |\Omega|$ operations.

The total cost of Algorithm 2 is bounded by

$$4 \times |\Omega|^2 \times |S| \; + \; 4 \times |\Omega| \times |S|^2 \; + \; 5 \times |\Omega| \times |S| \; + \; 6 \times |\Omega| \; + \; |S|$$

operations.

Parallel Testing of Regularity

We wish to eliminate the $|\Omega|^2 \times |S|$ term in the cost of the algorithm. This term is due to the calls to *trace* where the spanning tree may be of height $|\Omega|$. It is possible to combine the formation of the orbit and Schreier vector with the formation of all the z's, in effect reducing the cost of a call to *trace* to one operation. Of course, we need to store all the z's as they are being constructed, so our time saving comes at the cost of more space. This is presented as Algorithm 3.

Algorithm 3 : Parallel testing of regularity

```
Input : a set S of generators of a permutation group G on Ω;

Output : whether G is regular or not;

begin
  α := 1; orbit := {α};   (* α is orbit representative*)

  (*initialize information concerned with z's*)
  for each generator s of G do
    for each point β in Ω do used[s][β] := false; end for;
    z[s][α] := α^s;   used[s][α^s] := true;
  end for;

  (*construct orbit and z's at the same time*)
  for each point γ in orbit do
    for each generator g of G do

      if γ^g not in orbit then (*extend orbit*)
        add γ^g to orbit;

        (*determine one more image in each z*)
        for each generator s of G do
          image := z[s][γ]^g;
          if used[s][image] then
            exit with result "not regular";
          else
            z[s][γ^g] := image;   used[s][image] := true;
          end if;
        end for;

      end if;

    end for;
  end for;

  if orbit ≠ Ω then exit with result "not regular"; end if;
```

```
  for each generator s of G do
    if z[s] does not commute with each generator then
      exit with result "not regular";
    end if;
  end for;

  exit with result "regular";
end;
```

Consider the group of degree 8 generated by a=(1,2,3,4)(5,6,7,8) and b=(1,5)(2,8)(3,7)(4,6). Algorithm 3 determines $z[a]$ = (1,2,3,4)(5,8,7,6) and $z[b]$ = (1,5)(2,6)(3,7)(4,8), both of which commute with the generators a and b. Hence, the group is regular.

Consider the group of degree 4 generated by a=(1,2,3,4) and b=(1,4)(2,3). Algorithm 3 determines $z[a]$=(1,2,3,4) and $z[b]$ = (1,4,3,2). They are both permutations and both commute with the generator a, but neither of them commutes with the generator b. Hence, the group is not regular.

The analysis is straightforward. Algorithm 3 is an $O(|\Omega| \times |S|^2)$ algorithm.

Testing Semiregularity

Suppose that the group G is not transitive. Then G has orbits

$$\Delta_0, \Delta_1, \cdots \Delta_{m-1}$$

with orbit representatives

$$\delta_0, \delta_1, \cdots \delta_{m-1}.$$

If G is semiregular then

(1) the orbits $\Delta_0, \Delta_1, \cdots \Delta_{m-1}$ all have the same size,

(2) G acting on the orbit Δ_0 is regular, and

(3) there is a permutation z of Ω such that

$$\delta_0^z = \delta_1, \ \delta_1^z = \delta_2, \ \cdots \ \delta_{m-1}^z = \delta_0,$$

and such that z commutes with all the generators of G.

The previous algorithm determines the truth of (2), so we will only consider determining the truth of (3). Again, the conditions uniquely determine the action of z. It remains to check that z is a permutation and that z does indeed commute with all the generators. The algorithm is Algorithm 4.

Algorithm 4 : Testing semiregularity

Input : a set S of generators of a nontransitive group G acting on Ω;

Output : whether G is semiregular;

```
begin

  form the orbit representatives δ0, δ1, ..., δm-1
    and a Schreier vector v of the orbit Δ0;

  use Algorithm 3 to decide whether G is regular on Δ0;
  if G is not regular on Δ0 then
    exit with result "not semiregular";
  end if;

  (*construct z and check it is a permutation*)
  for each point β in Ω do used[β] := false; end for;

  for each point β in Δ0 do
    g := trace(β, v);
    for i := 0 to m-1 do
      image := δ^g_{i+1 mod m};
      if used[image] then
        exit with result "not semiregular";
      else
        used[image] := true;   z[δ^g_i] := image;
      end if;
    end for;
  end for;

  if z commutes with each generator of G then
    exit with result "semiregular";
  else
    exit with result "not semiregular";
  end if;

end;
```

Of course, we can determine z in parallel with constructing the orbit, and therefore effectively reduce the cost of the call to *trace* to one operation.

Consider the group of degree 8 generated by a=(1,2)(3,4)(5,6)(7,8) and b=(1,3)(2,4)(5,7)(6,8). The orbit representatives are 1 and 5. The group acting on Δ_0={1,2,3,4} is regular. The algorithm constructs z=(1,5)(2,6)(3,7)(4,8) which is a permutation and which commutes with both generators. Hence, the group is semiregular.

Summary

This chapter has presented efficient algorithms for deciding whether or not a permutation group, given by a set of generators, is regular or semiregular.

Exercises

(1/Easy) Write the modified version of *trace* mentioned in the analysis of Algorithm 2 and show that the cost of *trace* (β, v) is $3 \times d(\beta)$.

(2/Moderate) Modify Algorithm 4 to form z in parallel with constructing the orbit Δ_0, so that the effective cost of *trace* is one operation. Analyse the resulting algorithm.

Bibliographical Remarks

The algorithms to test regularity (and semiregularity) were described by Charles Sims in lectures given at Oxford in January and February 1973. The fundamental lemma can be proved using Proposition 4.3 in the book H. Wielandt, **Finite Permutation Groups**, Academic Press, New York, 1964.

Chapter 9. Primitivity

This chapter is concerned with partitioning the set Ω of points so that the group G acts on the partition. That is, can we partition Ω into disjoint subsets $B_1, B_2, \ldots, B_r$ so that for all i and for all $g \in G$, B_i^g is some other subset B_j (where j depends on i and g, of course)?

This is important because the group G can then be studied as a permutation group on $\{1,2,\ldots,r\}$ rather than as a permutation group on Ω. The smaller degree of the permutations should simplify the computations.

Definitions

A partition

$$\Omega = \{ B_1 \mid B_2 \mid \cdots \mid B_r \}$$

of Ω into disjoint subsets is *invariant* under G if the image of each subset B_i of the partition under any element of the group G is also a subset of the partition. That is, for all i and for all $g \in G$, there is a j such that $B_i^g = B_j$. Or, alternatively, each element of G permutes the subsets amongst themselves.

As an example, the group G acting on $\{1,2,\ldots,20\}$ and generated by

a=(1,2,3,4,5)(6,10,13,15,9)(7,11,14,8,12)(16,17,18,19,20) and
b=(1,2)(3,4)(7,11)(8,10)(9,12)(14,15)(16,17)(18,19)

has the following invariant partitions.

(1) $B_1=\{1\}, B_2=\{2\}, \ldots, B_{20}=\{20\}$

(2) $B_1=\{1,2,\ldots,20\}$

(3) $B_1=\{1,2,\ldots,5\}$ $B_2=\{6,7,\ldots,15\}$ $B_3=\{16,17,\ldots,20\}$

(4) $B_1=\{1,2,\ldots,15\}$ $B_2=\{16,17,\ldots,20\}$

(5) $B_1=\{1,16\}$ $B_2=\{2,17\}$ $B_3=\{3,18\}$ $B_4=\{4,19\}$ $B_5=\{5,20\}$ $B_6=\{6,7,\ldots,15\}$

A *trivial* invariant partition of a set Ω is either

the discrete partition where $B_i = \{i\}$, for all $i \in \Omega$, or

the complete partition where $B_1 = \Omega$, or

the partitions formed by the orbits of G, where each B_i is a (union of) orbits of G.

The only non-trivial partition in the above example is (5). The first is the discrete partition. The second is the complete partition. The third and fourth are formed by the orbits of G.

A group G acting on a set Ω is *primitive* if there is no non-trivial partition of Ω that is invariant under G. A group that has a non-trivial invariant partition is called *imprimitive*.

The above group is imprimitive.

Generally, the concept of primitivity is restricted to *transitive* groups. They are groups with only one orbit, the whole of Ω. In this case, all the subsets in an invariant partition are the same size.

Finest Invariant Partition

The algorithm we will present finds the finest invariant partition of Ω where one of the subsets contains the set $\{\omega_1, \omega_2\}$, for the given points $\omega_1 < \omega_2$. The algorithm requires a generating set of the group G. Using this algorithm and running through all such pair of points, we can determine whether there is a non-trivial invariant partition and, hence, whether G is primitive or imprimitive.

Let

$$partition(\omega_1, \omega_2) = \{B_1 \mid B_2 \mid \cdots \mid B_r\}$$

be the finest invariant partition of Ω where the points ω_1 and ω_2 are in the same subset of the partition. For any point α, let $B(\alpha)$ denote the subset of $partition(\omega_1, \omega_2)$ which contains α. This partition may also be represented as a function

$$f(\omega_1, \omega_2) : \Omega \rightarrow \Omega$$

where each point α is mapped to the smallest point in the subset $B(\alpha)$. Then, if α is the smallest point in a subset, $B(\alpha) = f(\omega_1, \omega_2)^{-1}(\alpha)$.

The algorithm will have a partition

$$f : \Omega \rightarrow \Omega$$

which is an approximation to $f(\omega_1, \omega_2)$. At all times f will be a refinement of $f(\omega_1, \omega_2)$, so that $f^{-1}(f(\alpha)) \subseteq B(\alpha)$, for all points α. Initially, we force ω_1 and ω_2 to be in the same subset. The approximation will converge to an invariant partition by merging subsets. It must converge to $f(\omega_1, \omega_2)$, since $f(\omega_1, \omega_2)$ is the finest such partition.

The only requirement we have for merging subsets of f is that the final result must be invariant under G. So we check whether the partition f is invariant and correct the situation whenever we discover it is not yet invariant. The partition is invariant if, for every pair of points β_1 and β_2 where $f(\beta_1) = f(\beta_2)$ and for every generator s of G, $f(\beta_1^s) = f(\beta_2^s)$. When we discover a pair where $f(\beta_1) = f(\beta_2)$ but not $f(\beta_1^s) = f(\beta_2^s)$ then we merge the subsets containing $f(\beta_1^s)$ and $f(\beta_2^s)$.

Figure 1 : Merging Subsets of the Partition

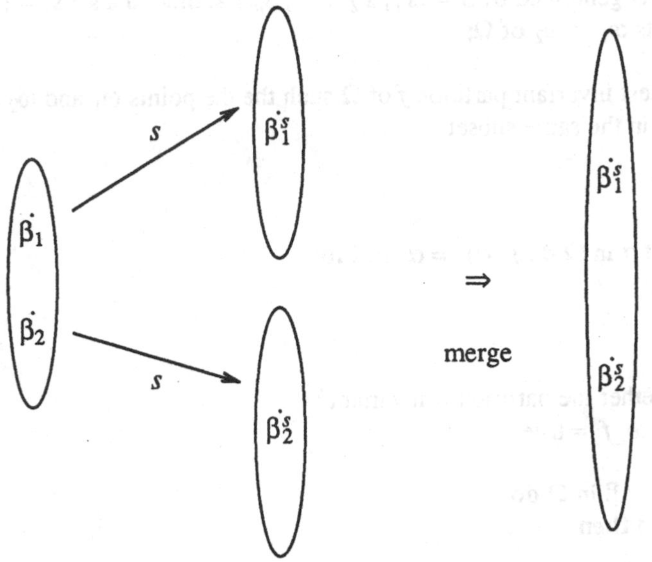

he procedure can be simplified greatly by noting that only pairs where β_1 is the smallest oint in the subset have to be used. Let α be the smallest point in the subset. If we verify that he generator s always maps such pairs to the same subset then for a general pair we have

$$f(\beta_1^s) = f(\alpha^s) = f(\beta_2^s)$$

ecause we have checked the action of s on the pairs α and β_i, for i=1,2. This gives us lgorithm 1.

Algorithm 1 : Minimal partition given a pair of points

```
Input : a group G generated by S = {s1, s2, . . . , sm} acting on a set Ω = {1,2,...,n};
    two points ω1 < ω2 of Ω;

Output : the finest invariant partition f of Ω such the the points ω1 and ω2
        are in the same subset;

begin

  for each point α in Ω do f (α) := α; end for;
  f (ω2) := ω1;

  repeat

    (*check whether the partition is invariant*)
    no_change_to_f := true;

    for each point β in Ω do
      if f (β) ≠ β then
        α := f (β);
        for each generator s of G do
          if f (α^s) ≠ f (β^s) then
            no_change_to_f := false;
            let δ be the smaller and let γ be the larger of f (α^s) and f (β^s);
            for each point τ in f^-1 (γ) do f (τ) := δ; end for;
          end if;
        end for;

      end if;
    end for;

  until no_change_to_f;

end;
```

Let us consider the group G of degree 20 that we have been using as an example. Let $\omega_1 = 3$ and $\omega_2 = 18$. The first iteration of the repeat loop in Algorithm 1 considers the pairs (18,3), (19,4), and (20,5) as (β,α) and consequently merges {19} with {4}, {20} with {5}, and {16} with {1}. The second iteration considers the pairs (16,1), (17,2), (18,3), (19,4), and (20,5) thus merging {17} with {2}. The third iteration considers the same pairs but has no merges. Therefore the loop terminates, giving the partition {1,16} {2,17} {3,18} {4,19} {5,20} {6} {7} {8} ... {14} {15}.

Eliminating Redundant Checking

The example of Algorithm 1 clearly demonstrates that each iteration of the **repeat** loop may check the same pair (β,α). In fact, the second time we check a pair (β,α) we know that each generator s maps α and β to the same subset. Since each pair is actually $(\beta, f(\beta))$, the only way for a new pair to be considered during the next iteration is if $f(\beta)$ is redefined. That is, if the subset containing $f(\beta)$ is merged with some other subset.

Suppose that $f(\beta) = \alpha$ is redefined to α'. So now $f(\alpha) = \alpha'$. Suppose we check that every generator maps α and α' to the same subset. We have previously checked that every generator maps β and α to the same subset. Hence, we can conclude that every generator maps β and α', the new $f(\beta)$, to the same subset.

The consequence of the above discussion is that we need only check the pairs (α, α') where α was the representative of a subset at the time the subset was merged with an earlier subset. (That is, α once played the role of γ (or ω_2) in Algorithm 1.) Once a point β is checked it has ceased to be a subset representative and will never be checked again.

To implement this improvement, we keep a set C of the points that have been subset representatives of merged subsets and have not yet been checked since losing their status as subset representative. Initially, ω_2 is the only point in the set. The improved algorithm is Algorithm 2.

Algorithm 2 : Minimal partition given a pair of points

```
Input : a group G generated by S = {s1, s2, . . . , sm} acting on a set Ω = {1,2,...,n};
        two points ω1 < ω2 of Ω;
Output : the finest invariant partition f of Ω such the the points ω1 and ω2
                are in the same subset;
begin
  for each point α in Ω do f (α) := α; end for;
  f (ω2) := ω1;  C := {ω2};

  repeat
    choose β ∈ C;  C := C - {β};  α := f (β);
    for each generator s of G do
      if f (α^s) ≠ f (β^s) then
        let δ be the smaller and let γ be the larger of f (α^s) and f (β^s);
        for each point τ in f^-1(γ) do f (τ) := δ; end for;
        C := C ∪ {γ};
      end if;
    end for;
  until C is empty;
end;
```

Repeating the previous example, we begin with $C = \{18\}$. The first value of β is 18 which merges $\{19\}$ with $\{4\}$ so that $C = \{19\}$. The second value of β is 19 which merges $\{20\}$ with $\{5\}$ so that $C = \{20\}$. The third value of β is 20 which merges $\{16\}$ with $\{1\}$ so that $C = \{16\}$. The fourth value of β is 16 which merges $\{17\}$ with $\{2\}$ so that $C = \{17\}$. The last value of β is 17 which causes no merging.

A point γ is added to C when a merging of subsets occurs and γ becomes an ex-representative of a subset, or when $\gamma = \omega_2$. Hence, each point (except ω_1) is added to C at most once. The number of points added to C is at most $|\Omega| - 1$. Therefore, there are at most $|\Omega| - 1$ iterations of the **repeat** loop, and at most $|\Omega| - 2$ occasions when $f(\alpha^s) \neq f(\beta^s)$. The merging of two subsets requires running through $f^{-1}(\gamma)$. One way to do this is to run through all the points and check their function value. This requires $|\Omega|$ operation to find the points τ, and at most $|\Omega| - 1$ operations to redefine all the $f(\tau)$, because the maximum size of a subset being merged is $|\Omega| - 1$. Once we add in the few miscellaneous costs the total worst case cost for Algorithm 2 becomes

$$2 \times |\Omega|^2 + 2 \times |S| \times \left[\, |\Omega| - 1 \, \right] - 3 \times |\Omega| + 4 \ \text{operations.}$$

Testing Primitivity

This section restricts attention to transitive groups. By doing this, we know that any non-trivial invariant partition has a pair $\{1, \omega\}$ of points contained in $B(1)$.

One way to test primitivity is to determine the set of all minimal partitions of Ω. If this set is empty, then the group G is primitive. An invariant partition is *minimal* if it has no refinement that is non-trivial and invariant.

Our assumption of the transitivity of G implies that each *partition* (ω_1, ω_2) is *partition* $(1, \omega)$, for some $\omega \neq 1$. Consider a fixed $\omega \neq 1$ and the corresponding *partition* $(1, \omega)$. For each $\omega' \in B(1)$, *partition* $(1, \omega')$ is a refinement of *partition* $(1, \omega)$. (They may be equal.) Hence, the set of minimal partitions is contained in

$$\{ \, \textit{partition}\,(1, \omega) \mid \omega \neq 1, \ \textit{partition}\,(1, \omega) \neq \Omega, \ B(1) \text{ does not contain } \omega',\ 1 < \omega' < \omega \, \}.$$

So an algorithm to test primitivity may use Algorithm 2 to construct *partition* $(1, \omega)$, for all $\omega \neq 1$. Those partitions that are complete are disregarded. The subset $B(1)$ may be checked to verify the last condition.

The last condition can also be checked within (a modified) Algorithm 2. There exists such an ω' if and only if Algorithm 2 merges $B(\delta)$ and $B(\gamma)$ where $\delta = 1 = \omega_1$ and $\gamma < \omega_2 = \omega$.

If we are only interested in a yes/no answer to the primitivity of the group then there is no need to construct the set of minimal partitions. The answer is no as soon as a *partition* $(1, \omega) \neq \Omega$ is constructed.

Summary

This chapter has presented an algorithm for finding partitions of the points Ω that are invariant under the action of the group. The generators of the group provided enough information for this algorithm, which finds the finest invariant partition having two given points in the same subset. The algorithm presented is $O(|\Omega|^2)$. The primitivity of a group can be determined using this algorithm in time $O(|\Omega|^3)$.

Exercises

(1/Easy) Determine whether the group of degree 8 generated by $a = (1,2,3,4)(5,6,7,8)$ and $b = (1,5)(2,8)(3,7)(4,6)$ is primitive or imprimitive.

(2/Easy) Determine whether the group of degree 4 generated by $a = (1,2,3,4)$ and $b = (1,4)(2,3)$ is primitive or imprimitive.

(3/Easy) Write out in detail the algorithm for testing primitivity. Assume the output is only a yes/no answer.

Bibliographical Remarks

This chapter has presented the work of M. D. Atkinson, "*An algorithm for finding the blocks of a permutation group*", Mathematics of Computation, **29**, 131 (1975) 911-913. The paper also considers representing the set C of Algorithm 2 as a circular list. However, the order of the worst case cost is not improved.

The problem is similar to the union-find problem. If one represents each subset of the partition as a list and merges the smaller subset with the larger subset then Algorithm 2 is $O(|\Omega| \times log(|\Omega|))$, and if one represents the subsets as trees and performs path compression then Algorithm 2 is essentially linear in $|\Omega|$ and $|S|$. However, in practice, a simple list is adequate. See G. Butler, "*An analysis of Atkinson's algorithm*", Basser Department of Computer Science Technical Report **259**, May 1985, for these analyses, and see M.D. Atkinson, R.A. Hassan, and M.P. Thorne, "*Group theory on a micro-computer*", in **Computational Group Theory**, M.D. Atkinson (ed.), Academic Press, New York, 1984, for an experimental comparison of these representations.

Chapter 10. Inductive Foundation

The foundation for the inductive algorithms which handle large permutation groups is a chain of stabilisers. A chain of stabilisers is represented by a base and strong generating set. The information required to go from one stabiliser to the next is contained in a Schreier vector for the stabiliser.

This chapter defines the above concepts and presents algorithms for the fundamental operations which use a base and strong generating set.

Definitions

Let G be a permutation group acting on the points Ω. Let $\beta \in \Omega$ be any point. We define the *stabiliser of* β *in* G by

$$G_\beta = \{ g \in G \mid \beta^g = \beta \},$$

the set of elements in G which fix or stabilise the point β. This set is a *subgroup*.

For example, in the symmetries of the square acting on {1,2,3,4} the stabiliser of the point 2 is G_2 = {*elt* [1]=/1 2 3 4/, *elt* [7]=/3 2 1 4/=(1,3)}.

The index $|G{:}G_\beta|$ of the stabiliser is the size of the orbit β^G. For each point $\gamma \in \beta^G$, the set of elements

$$\{ g \in G \mid \beta^g = \gamma \}$$

is a (right) coset of the stabiliser G_β in G. If v is a Schreier vector of the orbit β^G, then *trace* (γ, v) is a representative of this coset. Thus,

$$\{ \mathit{trace}(\gamma, v) \mid \gamma \in \beta^G \}$$

is a *set of coset representatives* of G_β in G.

For example, in the symmetries of the square {*elt* [4]=(1,2,3,4), *elt* [8]=(1,4)(2,3)} is the set of elements that map the point 2 to the point 3. Hence, it is a coset of G_2. As a set of coset representatives we could take {*elt* [1], *elt* [2], *elt* [3], *elt* [4]} since this set contains precisely one element mapping the point 2 to each point in {1,2,3,4}, the orbit of 2 under G.

We have demonstrated the one-to-one correspondence between the cosets of the stabiliser G_β in G and the points of the orbit β^G. Note that

$$\{ g \in G \mid \beta^g = \gamma \} = G_\beta \times \mathit{trace}(\gamma, v).$$

Since G_β is a subgroup, we can iterate the process of defining stabilisers and consider a stabiliser in G_β. That is, a subgroup of G of the elements which stabilise two points. In the general case, we define the stabiliser

$$
\begin{aligned}
G_{\beta_1, \beta_2, \ldots, \beta_i} &= \text{stabiliser of } \beta_i \text{ in } G_{\beta_1, \beta_2, \ldots, \beta_{i-1}} \\
&= \left[G_{\beta_1, \beta_2, \ldots, \beta_{i-1}} \right]_{\beta_i} \\
&= \{ g \in G_{\beta_1, \beta_2, \ldots, \beta_{i-1}} \mid g \text{ fixes } \beta_i \} \\
&= \{ g \in G \mid g \text{ fixes each of } \beta_1, \beta_2, \ldots, \beta_i \}.
\end{aligned}
$$

Hence, there is a one-to-one correspondence between the points in the orbit of β_i under $G_{\beta_1, \beta_2, \ldots, \beta_{i-1}}$ and the cosets of $G_{\beta_1, \beta_2, \ldots, \beta_i}$ in $G_{\beta_1, \beta_2, \ldots, \beta_{i-1}}$. That is,

$$
| G_{\beta_1, \beta_2, \ldots, \beta_{i-1}} : G_{\beta_1, \beta_2, \ldots, \beta_i} | = | \beta_i^{G_{\beta_1, \beta_2, \ldots, \beta_{i-1}}} |.
$$

If $v^{(i)}$ is a Schreier vector of the orbit then

$$
\{ \mathit{trace}(\gamma, v^{(i)}) \mid \gamma \in \beta_i^{G_{\beta_1, \beta_2, \ldots, \beta_{i-1}}} \}
$$

is a set of coset representatives of $G_{\beta_1, \beta_2, \ldots, \beta_i}$ in $G_{\beta_1, \beta_2, \ldots, \beta_{i-1}}$.

For example, in the symmetries of the square the stabiliser $G_{2,3}$ is the subgroup of G_2 which fixes 3. This is just the identity subgroup. The set of coset representatives of the identity subgroup in G_2 is $\{elt[1], elt[7]\}$. These elements map 3 to 3 and 1 respectively, and these two points are precisely the points in the orbit 3^{G_2}.

Eventually, we will stabilise so many points that the only element in the stabiliser is the identity element. This will give us a *chain of stabilisers*

$$
\{\mathit{identity}\} = G_{\beta_1, \beta_2, \ldots, \beta_k} \leq G_{\beta_1, \beta_2, \ldots, \beta_{k-1}} \leq \cdots \leq G_{\beta_1} \leq G.
$$

A sequence B of points $[\beta_1, \beta_2, \ldots, \beta_k]$ is called a *base* for G if the corresponding stabiliser chain does terminate with the identity subgroup. That is, the only element of the group G which fixes each of the points $\beta_1, \beta_2, \ldots, \beta_k$ is the identity.

For the example of the symmetries of the square we have shown that [2,3] is a base.

One useful property of a base is that an element g of the group G is uniquely determined by the sequence $\beta_1^g, \beta_2^g, \ldots, \beta_k^g$ (called the *base image* of g).

For example, the element $elt[6]$ of the symmetries of the square is the unique element which maps the base [2,3] to [1,4] respectively.

To "know" each of the stabilisers in the chain we at least need generators for each of them. A subset S of the group G is called a *strong generating set* of G relative to the base B if S contains generators for each stabiliser in the chain.

For example, the set $\{elt[2], elt[7]\}$ is a strong generating set of the symmetries of the square relative to the base [2,3].

Associated with a base and strong generating set are the various stabilisers, orbits, Schreier vectors, and sets of coset representatives. The notation we will use follows.

The stabilisers in the chain are denoted by

$$
G^{(i)} = G_{\beta_1, \beta_2, \ldots, \beta_{i-1}}
$$

for $1 \le i \le k+1$. So, in particular, $G^{(1)} = G$ and $G^{(k+1)} = \{identity\}$.

The stabiliser $G^{(i)}$ is generated by

$$S^{(i)} = S \cap G^{(i)}$$

for $1 \le i \le k+1$.

The *basic orbit* of the stabiliser is

$$\Delta^{(i)} = \beta_i^{G^{(i)}}$$

for $1 \le i \le k$. The sizes of the orbits are called the *basic indices*, since $|\Delta^{(i)}|$ is the index of $G^{(i+1)}$ in $G^{(i)}$.

The Schreier vector of the basic orbit $\Delta^{(i)}$ with respect to the generators $S^{(i)}$ is denoted $v^{(i)}$.

A fixed set of coset representatives of $G^{(i+1)}$ in $G^{(i)}$ is denoted $U^{(i)}$ and is called a *basic transversal*. For example, the set

$$\{ \; trace(\gamma, v^{(i)}) \mid \gamma \in \Delta^{(i)} \; \}.$$

Examples

This section illustrates the concepts of base and strong generating set through examples. The examples range from the most elementary to large imprimitive groups. Indeed, we have tried to include as wide a range as possible so as not to reinforce some erroneous notion of a "typical" base and strong generating set.

The first group is the symmetries of the square. The group has degree 4 and is generated by

$a=(1,2,3,4)$, and
$b=(2,4)$.

It has order $8=2^3$. A base for the group is

[1, 2]

and a strong generating set relative to this base is

$s_1=a$, and
$s_2=b$.

The stabilisers are

$G=G^{(1)}=\langle a, b\rangle$, and
$G_1=G^{(2)}=\langle b\rangle$

and the coset representatives may be taken to be

$U^{(1)}=\{id, a, a^2, a^3\}$, and
$U^{(2)}=\{id, b\}$.

The Schreier vectors are

	1	2	3	4
$v^{(1)}$	0	a	a	a
$v^{(2)}$	0	0	0	b

The second group is the symmetries of the projective plane of order two. The group has degree 7 and is generated by

$a=(1,2,4,5,7,3,6)$, and
$b=(2,4)(3,5)$.

It has order $168=2^3 \times 3 \times 7$. A base for the group is

$[1, 2, 4]$

and a strong generating set relative to this base is

$s_1=a$,
$s_2=b$,
$s_3=(4,5)(6,7)$, and
$s_4=(4,6)(5,7)$.

The stabilisers are

$G=G^{(1)}=<a, b>$,
$G_1=G^{(2)}=<b, s_3, s_4>$, and
$G_{1,2}=G^{(3)}=<s_3, s_4>$

and the coset representatives may be taken to be

$U^{(1)}=\{id, a, a^2, a^3, a^4, a^5, a^6\}$,
$U^{(2)}=\{id, b, b\times s_3, b\times s_4, b\times s_3\times b, b\times s_3\times s_4\}$, and
$U^{(3)}=\{id, s_3, s_4, s_3\times s_4\}$.

The Schreier vectors are

	1	2	3	4	5	6	7
$v^{(1)}$	0	a	a	a	a	a	a
$v^{(2)}$	0	0	b	b	s_3	s_4	s_4
$v^{(3)}$	0	0	0	0	s_3	s_4	s_4

The third group was discovered by the French mathematician Mathieu. The group has degree 11. A base for the group is

[11, 10, 1, 2]

and a strong generating set relative to this base is

s_1=(1,2,3)(4,5,6)(7,8,9),
s_2=(2,4,3,7)(5,6,9,8),
s_3=(2,5,3,9)(4,8,7,6),
s_4=(1,10)(4,5)(6,8)(7,9), and
s_5=(1,11)(4,6)(5,9)(7,8).

It has order 7,920=$2^4 \times 3^2 \times 5 \times 11$.

The Schreier vectors are

	1	2	3	4	5	6	7	8	9	10	11
$v^{(1)}$	s_5	s_1	s_1	s_2	s_3	s_1	s_2	s_3	s_3	s_4	0
$v^{(2)}$	s_4	s_1	s_1	s_2	s_3	s_1	s_2	s_3	s_3	0	0
$v^{(3)}$	0	s_1	s_1	s_2	s_3	s_1	s_2	s_3	s_3	0	0
$v^{(4)}$	0	0	s_2	s_2	s_3	s_2	s_2	s_3	s_3	0	0

The fourth group has degree 21 and is generated by

a=(1,8,9)(2,11,15)(3,10,12)(4,14,19)(5,16,17)(6,21,20)(7,13,18),
b=(9,18,20)(12,19,17), and
c=(10,21,11)(13,16,14).

It has order 27,783=$3^4 \times 7^3$. A base for the group is

[1, 9, 8, 10, 2, 12]

and a strong generating set relative to this base is

$s_1=a$,
$s_2=b$,
$s_3=c$,
s_4=(8,13,21)(10,14,16),
s_5=(2,6,3)(4,5,7), and
s_6=(12,20,15)(17,19,18).

The Schreier vectors are

	1	2	3	4	5	6	7	8	9	10	11
$v^{(1)}$	0	a	s_5	a	a	s_5	a	a	a	s_4	c
$v^{(2)}$	0	0	0	0	0	0	0	0	0	0	0
$v^{(3)}$	0	0	0	0	0	0	0	0	0	s_4	c
$v^{(4)}$	0	0	0	0	0	0	0	0	0	0	c
$v^{(5)}$	0	0	s_5	0	0	s_5	0	0	0	0	0
$v^{(6)}$	0	0	0	0	0	0	0	0	0	0	0

12	13	14	15	16	17	18	19	20	21
b	s_4	c	s_6	c	s_6	b	s_6	b	s_4
b	0	0	s_6	0	s_6	b	s_6	b	0
0	s_4	c	0	c	0	0	0	0	s_4
0	0	0	0	0	0	0	0	0	c
0	0	0	0	0	0	0	0	0	0
0	0	0	s_6	0	0	0	0	s_6	0

The fifth group is the set of all operations of Rubik's cube. The group has degree 48 and is generated by

a=(1,3,8,6)(2,5,7,4)(9,48,15,12)(10,47,16,13)(11,46,17,14),
b=(6,15,35,26)(7,22,34,19)(8,30,33,11)(12,14,29,27)(13,21,28,20),
c=(1,12,33,41)(4,20,36,44)(6,27,38,46)(9,11,26,24)(10,19,25,18),
d=(1,24,40,17)(2,18,39,23)(3,9,38,32)(41,43,48,46)(42,45,47,44),
e=(3,43,35,14)(5,45,37,21)(8,48,40,29)(15,17,32,30)(16,23,31,22), and
f=(24,27,30,43)(25,28,31,42)(26,29,32,41)(33,35,40,38)(34,37,39,36).

The group has order

$$2^{27}3^{14}5^{3}7^{2}11 = 43\,252\,003\,274\,489\,856\,000.$$

A base for the group is

[1, 6, 3, 8, 21, 23, 26, 5, 29, 19, 7, 24, 25, 28, 31, 18, 4, 2]

and a strong generating set relative to this base is

$s_1=a$,
$s_2=b$,
$s_3=e$,
$s_4=(5,37,28,21)(8,26,29,32)(14,33,35,40)(15,27,30,43)(16,31,34,22)$,
$s_5=(5,45,13,37,21)(7,31,22,16,23)(26,27,33)(29,35,30)$,
$s_6=(19,23,34)(20,45,28)$,
$s_7=f$,
$s_8=(5,28,37)(16,34,31)$,
$s_9=(2,28,47,34)(24,41,38)(25,39,31,36,42,37)(29,43,30,40,35,32)$,
$s_{10}=(19,31,34)(20,37,28)$,
$s_{11}=(7,31,34)(13,37,28)$,
$s_{12}=(24,41,38)(32,43,40)$,
$s_{13}=(24,40)(25,39,34,37)(28,31,36,42)(32,38)(41,43)$,
$s_{14}=(25,28,31)(34,37,36)$,
$s_{15}=(2,31,34)(28,47,37)$,
$s_{16}=(2,31,39)(37,42,47)$,
$s_{17}=(2,18,39)(42,47,44)$,
$s_{18}=(2,10,39)(4,42,47)$, and
$s_{19}=(2,47)(39,42)$.

The sixth group has degree 14 and is generated by

$a=(1,2)(3,4)(5,6)(7,8)(9,10)(11,12)$, and
$b=(1,13)(2,3,7,5)(6,9,11,8)(10,14)$.

The group is imprimitive. It has order $10{,}752 = 2^9 \times 3 \times 7$. A base for the group is

$[1, 2, 3, 4, 7, 5]$

and a strong generating set relative to this base is

$s_1=a$,
$s_2=(2,7)(3,5)(6,11)(8,9)$,
$s_3=(3,7,11,8)(4,14)(5,6)(12,13)$,
$s_4=(3,13,11,14)(4,8)(5,6)(7,12)$,
$s_5=(4,12)(13,14)$,
$s_6=(7,8)(13,14)$, and
$s_7=(5,6)(13,14)$.

The Schreier vectors are

	1	2	3	4	5	6	7	8	9	10	11	12	13	14
$v^{(1)}$	0	a	s_3	s_3	s_2	s_7	s_2	s_6	s_2	a	s_4	s_3	s_4	s_4
$v^{(2)}$	0	0	s_3	s_3	s_2	s_7	s_2	s_6	s_2	0	s_4	s_3	s_4	s_4
$v^{(3)}$	0	0	0	s_3	0	0	s_3	s_6	0	0	s_4	s_3	s_4	s_4
$v^{(4)}$	0	0	0	0	0	0	0	0	0	0	0	s_5	0	0
$v^{(5)}$	0	0	0	0	0	0	0	s_6	0	0	0	0	0	0
$v^{(6)}$	0	0	0	0	0	s_7	0	0	0	0	0	0	0	0

The seventh group has degree 16 and is generated by

a=(1,2)(3,4)(5,6)(7,8)(9,10)(11,12)(13,14)(15,16), and
b=(1,2,5,3)(4,7)(6,9)(8,11)(10,13,16,15)(12,14).

The group is imprimitive. It has order $21{,}504 = 2^{10} \times 3 \times 7$. A base for the group is

[1, 2, 3, 4]

and a strong generating set relative to this base is

s_1=a,
s_2=(3,14,10,6)(4,5,9,13)(7,11)(8,12),
s_3=(2,6,10,14,11,7,3)(4,8,12,13,15,9,5),
s_4=(4,12)(6,11)(7,14)(8,9), and
s_5=(4,8)(6,7)(9,12)(11,14).

The eighth group has degree 18 and is generated by

a=(1,2)(3,4)(5,6)(7,8)(9,10)(11,12)(13,14)(15,16)(17,18), and
b=(1,2,5,3)(4,7)(6,9,12,11)(8,13,16,15)(10,14,18,17).

The group is imprimitive. It has order 508,032 = $2^7 \times 3^4 \times 7^2$. A base for the group is

[1, 2, 3, 4, 5, 6]

and a strong generating set relative to this base is

s_1=a,
s_2=(2,10)(3,6)(4,11)(5,17)(7,18)(8,14)(9,16)(12,13),
s_3=(3,15,7,10,6,18,13)(4,8)(5,9)(11,14)(16,17),
s_4=(4,16,5,8,14,11,17),
s_5=(5,14,8,16,11,17,9), and
s_6=(6,15,10,7,18,13,12).

The ninth group has degree 16 and is generated by

a=(1,15,7,5,12)(2,9,13,14,8)(3,6,10,11,4),
b=(1,7)(2,11)(3,12)(4,13)(5,10)(8,14), and
c=(1,16)(2,3)(4,5)(6,7)(8,9)(10,11)(12,13)(14,15).

The group is primitive. It has order 11,520=$2^8 \times 3^3 \times 5$. A base for the group is

[1, 2, 8, 3, 4]

and a strong generating set relative to this base is

s_1=a,
s_2=(2,4,6)(3,5,7)(8,9)(10,13,14,11,12,15),
s_3=(8,9)(10,11)(12,13)(14,15),
s_4=(3,12,14)(4,10,6,7,9,5)(8,11)(13,16,15),
s_5=(3,12,14)(4,9,6)(5,7,10)(13,16,15), and
s_6=(4,6)(5,7)(12,14)(13,15).

The tenth group has degree 31 and is generated by

a=(4,5)(6,8)(7,9)(10,12)(13,16)(15,19)(18,22)(20,24)
(23,27)(25,29)(26,30)(28,31), and
b=(1,2,3,4,6)(5,7,10,13,17)(8,11,14,18,23)
(9,12,15,20,25)(16,21,22,26,29)(19,24,28,31,27).

The group is primitive. It has order 9,999,360=$2^{10} \times 3^2 \times 5 \times 7 \times 31$. A base for the group is

[4, 2, 3, 1, 6]

and a strong generating set relative to this base is

$s_1=b$,
$s_2=(2,13)(3,18)(5,25)(7,26)(9,31)(11,16)(12,27)(14,28)(15,22)$
$(17,20)(19,24)(21,30)$,
$s_3=(1,8)(3,30,7,24)(9,28,17,22)(10,29)(11,16)(12,27)$
$(14,31,15,20)(18,19,26,21)$,
$s_4=(1,29)(8,10)(9,20)(11,27)(12,16)(17,31)(19,24)(21,30)$,
$s_5=(6,8)(7,9)(13,16)(15,19)(23,29)(25,27)(26,31)(28,30)$,
$s_6=(6,29)(7,31)(8,23)(9,26)(13,27)(15,30)(16,25)(19,28)$,
$s_7=(6,16)(7,19)(8,13)(9,15)(23,27)(25,29)(26,30)(28,31)$, and
$s_8=(6,19)(7,16)(8,15)(9,13)(23,30)(25,31)(26,27)(28,29)$.

Representing Elements and Testing Membership

Let's look at some consequences of knowing a base and strong generating set (and the Schreier vectors).

Recall that the cosets of a subgroup partition a group. In our case, the subgroup is a stabiliser G_{β_1} of the group G. This partition means that each element of the group can be written as a product

$$h \times u$$

where u is a coset representative, and h is an element of the subgroup. Each product gives a distinct element of the group (provided we use a fixed set of coset representatives). In this case, it says that each element of the group G is uniquely represented as a product

$$h \times u_1$$

where $h \in G^{(2)}$ and $u_1 \in U^{(1)}$. Using induction, we see that each element of G is uniquely represented as a product

$$u_k \times u_{k-1} \times \cdots \times u_1 \tag{4.1}$$

where $u_i \in U^{(i)}$, for $1 \leq i \leq k$, the length of the base.

With the correspondence between the coset representatives and the orbits, we know that $|U^{(i)}| = |\Delta^{(i)}|$. Hence,

$$|G| = \prod_{i=1}^{k} |U^{(i)}| = \prod_{i=1}^{k} |\Delta^{(i)}|.$$

Not only do we get the order of the group, but (4.1) tells us how to get each and every element of the group. Note that the base image of the element in (4.1) is

$$\beta_1^{u_1}$$
$$\beta_2^{u_2 \times u_1}$$
$$\beta_3^{u_3 \times u_2 \times u_1}$$
$$\cdots$$
$$\beta_i^{u_i \times u_{i-1} \times \cdots \times u_1}$$
$$\cdots$$
$$\beta_k^{u_k \times u_{k-1} \times \cdots \times u_1}$$

since the element u_j fixes $\beta_1, \beta_2, ..., \beta_{j-1}$. This is a useful relationship between the unique representation of an element by its base image and the unique representation as a product (4.1) of coset representatives, especially when the coset representatives come from the Schreier vectors. We will use these representations of elements heavily in the next few algorithms. The algorithms solve some elementary but necessary tasks.

The first task is to test membership of an arbitrary permutation on Ω in the group G. Algorithm 1 attempts to find the coset representatives involved in the product (4.1). If the permutation is an element then the coset representatives will be found. If the permutation is not an element of the group then it will not be possible to find the coset representatives.

Algorithm 1 : Testing Membership

Input : a group G acting on $\Omega = \{1,2,...,n\}$;
  a permutation g of $\Omega = \{1,2,...,n\}$;
  a base and strong generating set for G;
  sets $U^{(i)}, 1 \le i \le k$ of coset representatives for stabiliser chain;

Output : a boolean value *answer*, indicating whether $g \in G$;

function *is_in_group*(p : permutation; i : 1..k+1) : boolean;
(* return *true* if the permutation p is in the group $G^{(i)}$ *)
begin
  if $i = k+1$ **then**
    result is $p = id$;
  else
    find $u_i \in U^{(i)}$ such that $\beta_i^{u_i} = \beta_i^{p}$;
    if no such u_i exists **then**
      result is *false*;
    else
      result is *is_in_group*($p \times u_i^{-1}$, i+1);
    end if;
  end if;
end;

begin
  answer := *is_in_group*(g, 1);
end;

Algorithm 2 solves the same problem using the Schreier vectors to determine the coset representative.

Algorithm 2 : Testing Membership

Input : a group G acting on $\Omega = \{1,2,...,n\}$;
a permutation g of $\Omega = \{1,2,...,n\}$;
a base and strong generating set for G;
Schreier vectors $v^{(i)}, 1 \le i \le k$, for the stabiliser chain;

Output : a boolean value *answer*, indicating whether $g \in G$;

```
function is_in_group( p : permutation; i : 1..k+1 ) : boolean;
(* return true if the permutation p is in the group G^(i) *)
begin
  if i = k+1 then
    result is p = id;
  else
    if β_i^p ∈ Δ^(i) then
      result is is_in_group( p × trace ( β_i^p, v^(i) )^-1, i+1 );
    else
      result is false;
    end if;
  end if;
end;

begin
  answer := is_in_group( g, 1 );
end;
```

To analyse Algorithm 2, we note that the worst case cost of tracing $v^{(i)}$ is $2 \times |\Delta^{(i)}| \left[|\Omega|+1 \right]$ operations. We must also consider the $|\Omega|$ operations to test $p = id$; the operation to form the image of β_i under p; and the one operation to test membership of a point in $\Delta^{(i)}$. The total worst case cost is therefore

$$|\Omega| \times \left[1 + 2 \times k + 2 \times \sum_{i=1}^{k} |\Delta^{(i)}| \right] + 2 \times k + 2 \times \sum_{i=1}^{k} |\Delta^{(i)}| \text{ operations.}$$

Algorithm 3 presents a non-recursive version of Algorithm 2. Similarly a non-recursive version of Algorithm 1 could be devised.

Algorithm 3 : Testing Membership

Input : a group G acting on $\Omega = \{1,2,...,n\}$;
a permutation g on $\Omega = \{1,2,...,n\}$;
a base and strong generating set for G;
Schreier vectors $v^{(i)}, 1 \le i \le k$, for the stabiliser chain;

Output : a boolean value *answer*, indicating whether $g \in G$;

function *is_in_group*(p : permutation; i : 1..k+1) : boolean;
(* return *true* if the permutation p is in the group $G^{(i)}$ *)
begin
  for $j := i$ **to** k **do**
    if $\beta_j^{\,p \times u_i^{-1} \times u_{i+1}^{-1} \times \cdots \times u_{j-1}^{-1}} \in \Delta^{(j)}$ **then**
      $u_j :=$ *trace*($\beta_j^{\,p \times u_i^{-1} \times u_{i+1}^{-1} \times \cdots \times u_{j-1}^{-1}}, v^{(j)}$);
    else
      result is *false*;
    end if;
  end for;
  result is $p = u_k \times u_{k-1} \times \cdots \times u_i$;
end;

begin
  answer := *is_in_group*(g, 1);
end;

We will consider some examples of testing membership. Let G be the symmetries of the projective plane of order two, as given earlier. Let p=(1,2,3,4,5,6,7). We wish to know whether $p \in G$? Using Algorithm 1, we find $u_1 = a$ is a coset representative that maps 1 to 2, which is the image of 2 under p. Therefore, we ask is $p \times u_1^{-1}$=(2,7,6,5,3) $\in G^{(2)}$? We find $u_2 = b \times s_3 \times s_4$ is a coset representative that maps 2 to 7, which is the image of 2 under $p \times u_1^{-1}$. Therefore, we ask is $p \times u_1^{-1} \times u_2^{-1}$=(3,4,7) $\in G^{(3)}$? We find $u_3 = s_3 \times s_4$ is a coset representative that maps 4 to 7, which is the image of 4 under $p \times u_1^{-1} \times u_2^{-1}$. Therefore, we ask is $p \times u_1^{-1} \times u_2^{-1} \times u_3^{-1}$ = (3,7)(5,6) $\in G^{(4)} = \{id\}$? The answer to this is clearly no, so $p \notin G$.

Let us consider the same group, but take p=(1,4,2,3,7,5,6). We will use Algorithm 2 to decide whether $p \in G$. We find that the image of 1 under p is 4, which is in $\Delta^{(1)}$, and that $trace(4, v^{(1)}) = a^2$. Therefore, we ask if $p \times trace(4, v^{(1)})^{-1}$ = (2,5,7)(3,4,6) $\in G^{(2)}$? We find that the image of 2 under $p \times trace(4, v^{(1)})^{-1}$ is 5, which is in $\Delta^{(2)}$, and that $trace(5, v^{(2)}) = b \times s_3$. Therefore, we ask if $p \times trace(4, v^{(1)})^{-1} \times trace(5, v^{(2)})^{-1}$ = (4,7)(5,6) $\in G^{(3)}$? We find that the image of 4 under $p \times trace(4, v^{(1)})^{-1} \times trace(5, v^{(2)})^{-1}$ is 7, which is in $\Delta^{(3)}$, and that $trace(7, v^{(3)}) = s_3 \times s_4$. Therefore, we ask if $p \times trace(4, v^{(1)})^{-1} \times trace(5, v^{(2)})^{-1} \times trace(7, v^{(3)})^{-1} = id \in G^{(4)}$? The answer is clearly yes, so $p \in G$. Indeed, it tells us that $p = (s_3 \times s_4) \times (b \times s_3) \times (a^2)$.

The next task we wish to perform is to write an element of the group G as a product of strong generators. Again, we will attempt to write the element as a product of coset representatives. The Schreier vectors determine coset representatives as products of strong generators (and their inverses), so we will use the Schreier vectors. The procedure *trace* will be assumed to also return the element as a product of the strong generators. Since an element is determined by its base image, we will represent some elements in the algorithm by a base image $[\gamma_1, \gamma_2, \ldots, \gamma_k]$. The algorithm to perform the task is Algorithm 4.

Algorithm 4 : Element as Product of Strong Generators

Input : a group G acting on $\Omega = \{1,2,\ldots,n\}$;
    an element g of G;
    a base and strong generating set for G;
    Schreier vectors $v^{(i)}$, $1 \le i \le k$, for the stabiliser chain;

Output : a symbolic product, *word*, expressing g in terms of the strong generators;
begin
    $[\gamma_1, \gamma_2, \ldots, \gamma_k] := [\beta_1, \beta_2, \ldots, \beta_k]^g$;
    $word := \varepsilon$;   (*the empty word*)
    for $i := 1$ **to** k **do**
        $u_i := trace(\gamma_i, v^{(i)})$;   (*$\gamma_i = \beta_i^{g \times u_1^{-1} \times u_2^{-1} \times \cdots \times u_{i-1}^{-1}}$*)
        $word := u_i \times word$;   (*regarding u_i as a word in $S^{(i)}$*)
        $[\gamma_1, \gamma_2, \ldots, \gamma_k] := [\gamma_1, \gamma_2, \ldots, \gamma_k]^{u_i^{-1}}$;
    end for;
end;

We will now consider an example of Algorithm 4. Let G be the sixth group in our examples. We know that

$$g = a \times b = (1,3,4,7,6,2,13)(5,9,14,10,11,12,8)$$

is an element of the group. Using Algorithm 4, we will express this element as a product of the strong generators. The base image of g is Figure 1.

Figure 1

	$\xrightarrow{g}$	
1		3
2		13
3		4
4		7
7		6
5		9

The element $trace(3, v^{(1)})$ is $a \times s_2 \times s_3$. The resulting base image $[\gamma_1, \gamma_2, \ldots, \gamma_k]$ is Figure 2.

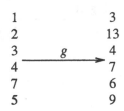

Figure 2

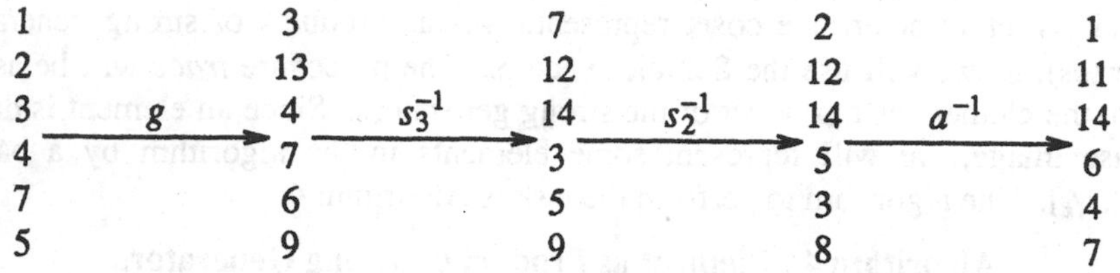

The element *trace* (11, $v^{(2)}$) is $s_2 \times s_3 \times s_4 \times s_4$. The resulting base image $[\gamma_1, \gamma_2, ..., \gamma_k]$ is Figure 3.

Figure 3

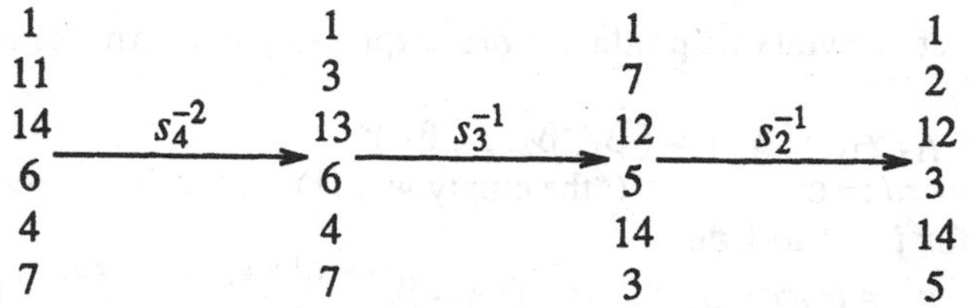

The element *trace* (12, $v^{(3)}$) is $s_4 \times s_3$. The resulting base image $[\gamma_1, \gamma_2, ..., \gamma_k]$ is Figure 4.

Figure 4

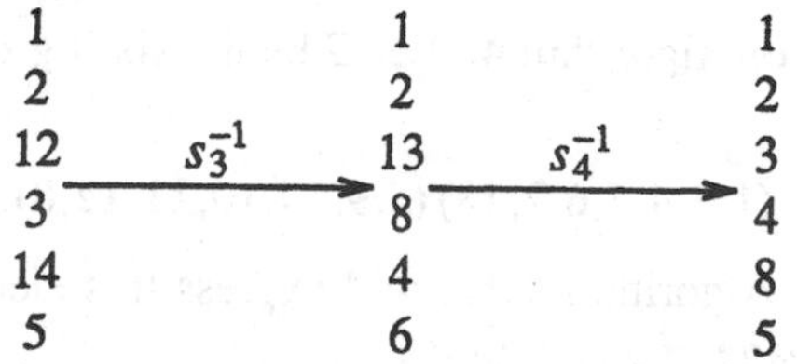

Therefore, *trace* (4, $v^{(4)}$) is the identity, leaving the base image unchanged. The element *trace* (8, $v^{(5)}$) is s_6, giving the base image in Figure 5.

Figure 5

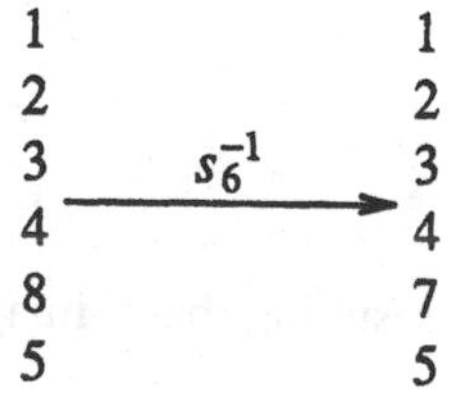

Therefore, *trace*(5, $v^{(6)}$) is the identity, leaving the base image unchanged. Hence, the result is

$$g = (s_6) \times (s_4 \times s_3) \times (s_2 \times s_3 \times s_4^2) \times (a \times s_2 \times s_3).$$

Note, that the above example suggests an improvement in Algorithm 4. At no stage did we need the complete element *trace*(γ_i, $v^{(i)}$). We simply needed the coset representative as a word, and we needed its action on a certain sequence of points. This action could be calculated from the strong generators (and their inverses) as we determine the word. Thus, a major improvement to Algorithm 4 would be to modify *trace* to return a word, and the image of a sequence of points (rather than an element).

Note also that $\gamma_1, \gamma_2, \ldots, \gamma_{i-1}$ are fixed by *trace*(γ_i, $v^{(i)}$), since these are the base points. This fact could also be used to improve Algorithm 4.

The last task we wish to consider here is to determine the element with a given base image. Since the base image uniquely determines the element, this is a reasonable request. Again, we will use the Schreier vectors to determine the coset representatives. The algorithm is given as Algorithm 5.

Algorithm 5 : Element from a Base Image

Input : a group G acting on $\Omega = \{1,2,\ldots,n\}$;
  the base image $[\gamma_1, \gamma_2, \ldots, \gamma_k[$ of an element g of G;
  a base and strong generating set for G;
  Schreier vectors $v^{(i)}$, $1 \le i \le k$, for the stabiliser chain;

Output : the element $g = u_k \times u_{k-1} \times \cdots \times u_1$;

```
begin
  g := id;
  for i := 1 to k do
    u_i := trace( γ_i, v^(i) );
    g := u_i × g;
    [ γ_1, γ_2, ..., γ_k ] := [ γ_1, γ_2, ..., γ_k ]^(u_i^-1) ;
  end for;
end;
```

Enumerating All Elements

In this section we present some straightforward algorithms for enumerating all the elements of a permutation group. We will consider two representations of an element - as a complete permutation, and as a base image. Both rely on essentially enumerating all the products of coset representatives.

Algorithm 6 sketches a non-recursive approach to enumerating all the elements as permutations. Later we will present a more precise recursive version.

Algorithm 6 : Enumerating All Elements

Input : a permutation group G;
a base and strong generating set for G;
the sets $U^{(i)}$, $1 \le i \le k$, of coset representatives of the stabiliser chain;

Output : the elements of the group G (without repetition);

begin

for each $u_1 \in U^{(1)}$ **do**
for each $u_2 \in U^{(2)}$ **do**

.
.
.

for each $u_k \in U^{(k)}$ **do**

$g := u_k \times u_{k-1} \times \cdots \times u_1$; (*next element of G*)

end for;

.
.
.

end for;
end for;

end;

Provided the identity is the "first" coset representative in each $U^{(i)}$, then Algorithm 6 enumerates the elements of $G^{(k)}$, then the elements of $G^{(k-1)} - G^{(k)}$, and so on until at last it enumerates the elements of $G^{(1)} - G^{(2)}$.

The next algorithm enumerates the base images of the elements. It is important to note that in the base image of $u_k \times u_{k-1} \times \cdots \times u_1$, the i-th point γ_i of the image is

$$\beta_i^{u_i \times u_{i-1} \times \cdots \times u_1}$$

The point $\beta_i^{u_i}$ runs over $\Delta^{(i)}$ as u_i varies, so the image point γ_i runs over

$$\left[\Delta^{(i)}\right]^{u_{i-1} \times u_{i-2} \times \cdots \times u_1}.$$

With this explanation, we are ready to present Algorithm 7.

Algorithm 7 : Enumerating All Base Images

Input : a permutation group G;
a base and strong generating set for G;
the Schreier vectors $v^{(i)}$, $1 \le i \le k$, of the stabiliser chain;

Output : the base images of the elements of the group G (without repetition);

begin

for each $\gamma_1 \in \Delta^{(1)}$ **do**

$u_1 := trace(\gamma_1, v^{(1)})$;

for each $\gamma_2 \in \left[\Delta^{(2)}\right]^{u_1}$ **do**

$u_2 := trace(\gamma_2, v^{(2)})$;

.

.

.

for each $\gamma_k \in \left[\Delta^{(k)}\right]^{u_{k-1} \times u_{k-2} \times \cdots \times u_1}$ **do**

$image := [\gamma_1, \gamma_2, ..., \gamma_k]$; (*next base image*)

end for;

.

.

.

end for;

end for;

end;

These algorithms are perhaps more clearly presented recursively. This is done in Algorithm 8, which enumerates the base images, but is easily modified to enumerate the complete permutations.

Algorithm 8 : Enumerating All Base Images

Input : a permutation group G;
a base and strong generating set for G;
the Schreier vectors $v^{(i)}$, $1 \le i \le k$, of the stabiliser chain;

Output : the base images of the elements of the group G (without repetition);

procedure *base_image*($i : 1..k+1$;

$\quad\quad\quad\quad\quad$ [$\gamma_1, \gamma_2, ..., \gamma_{i-1}$] : sequence of points;

$\quad\quad\quad\quad\quad$ $u_{i-1} \times u_{i-2} \times \cdots \times u_1$: element $\quad\quad\quad$);

(*

Enumerate all the base images which begin with [$\gamma_1, \gamma_2, ..., \gamma_{i-1}$],

where $u_{i-1} \times u_{i-2} \times \cdots \times u_1$ is an element mapping

the initial segment of the base to the given sequence.

*)

begin

$\quad$ **if** $i = k+1$ **then**

$\quad\quad$ *image* := [$\gamma_1, \gamma_2, ..., \gamma_{i-1}$]; $\quad$ (*next base image*)

$\quad$ **else**

$\quad\quad$ $g := u_{i-1} \times u_{i-2} \times \cdots \times u_1$;

$\quad$ **for** each $\gamma_i \in \left[\Delta^{(i)} \right]^g$ **do**

$\quad\quad$ $u_i := trace\,(\gamma_i{}^{g^{-1}}, v^{(i)}\,)$;

$\quad\quad$ *base_image*($i+1$, [$\gamma_1, \gamma_2, ..., \gamma_i$], $u_i \times g$);

$\quad$ **end for**;

$\quad$ **end if**;

end;

begin

$\quad$ *base_image*(1, empty sequence, *id*);

end;

Summary

This chapter has introduced the stabiliser chain, the inductive foundation of the powerful algorithms for handling large permutation groups. The representation of elements by their base image, and as a product of coset representatives is central to the power of these algorithms. They are applied here to the performance of the tasks of testing membership, writing elements as products of the strong generators, and the enumeration of all the elements of a group.

We have presented several examples. It is hoped the reader will use them in the study of these and later algorithms.

Exercises

(1/Easy) Give a base and strong generating set for the symmetric group of degree 4.

(2/Moderate) Let $G=<s>$ be a cyclic group. Let $B = [\beta_1, \beta_2, ..., \beta_k]$ be a base for G. Devise an algorithm that determines a strong generating set of G relative to B.

(3/Difficult) Construct (perhaps using the definition of stabiliser and automorphism) a base and strong generating set of the automorphism group of Petersen's graph.

(4/Easy) Construct Schreier vectors for some, or all, of the examples for which they are not given.

(5/Moderate) Modify *trace* to return both a word in the strong generators for the coset representative, and the image under the coset representative of a sequence of points. The sequence of points is a parameter. Do not return an element.

Do the same for the inverse of the coset representative.

(6/Moderate) Modify Algorithm 5 so that, given a sequence $[\gamma_1, \gamma_2, ..., \gamma_{i-1}]$ of points, it determines whether or not there is an element of the group whose base image begins with $[\gamma_1, \gamma_2, ..., \gamma_{i-1}]$.

Bibliographical Remarks

The concept of a stabiliser, and the one-to-one correspondence between the cosets of a stabiliser and the orbit of the point being stabilised are old. They are discussed in H. Wielandt, **Finite Permutation Groups**, Academic Press, New York, 1964.

The use of a stabiliser chain and the sets $U^{(i)}$ of coset representatives to represent a permutation group are presented in C. C. Sims, "*Computational methods in the study of permutation groups*", **Computational Problems in Abstract Algebra**, (Proceedings of a conference, Oxford, 1967), John Leech (editor), Pergamon, Oxford, 1970, 169-183. This paper also notes the unique representation of an element as a product of coset representatives, even though the base is assumed to be $[1,2,...,|\Omega|-1]$. The representation is used to test membership similarly to Algorithm 1.

The terms *base* and *strong generating set* are due to C. C. Sims, "*Determining the conjugacy classes of a permutation group*", **Computers in Algebra and Number Theory** (Proceedings of the Symposium on Applied Mathematics, New York, 1970), G. Birkhoff and M. Hall, Jr (editors), SIAM-AMS Proceedings, volume 4, American Mathematics Society, Providence, Rhode Island, 1971, 191-195; and C. C. Sims, "*Computation with permutation groups*", (Proceedings of the Second Symposium on Symbolic and Algebraic Manipulation, Los Angeles, 1971), S. R. Petrick (editor), Association of Computing Machinery, New York, 1971, 23-28. In these papers, a base is not necessarily $[1,2,...,|\Omega|-1]$; the representation of an element by its base image is noted; and the representation of an element as a product of coset representatives is generalized to its present form.

The use of Schreier vectors in testing membership comes from the Oxford lectures of Sims in January and February 1973.

Chapter 11. Backtrack Search

In this chapter we consider searching the elements of a permutation group G with a base and strong generating set. The search aims to find a base and strong generating set for the subgroup of elements which satisfy a given property P. The search is through the tree of all base images of elements of G. Heuristic methods prune this tree. Some heuristics are *problem-dependent,* but others are applicable to all properties P, and are called *problem-independent.*

The backtrack search will enable us to construct a base and strong generating set for centralizers, normalizers, intersections, set stabilisers, and the automorphism group of a group.

Strong Generators of the Subgroup

Let P be a property of (some of) the elements of G. We require that P be decidable, so that given an element $g \in G$ it can be determined whether g satisfies property P or not. Furthermore, we require that the elements of G that satisfy the property P form a subgroup of G. We call this subgroup $H(P)$.

$$H(P) = \{ g \in G \mid g \text{ satisfies } P \}.$$

The first (and most important) consideration is that the algorithm construct a base and strong generating set of the subgroup, since they provide enough information to further investigate the subgroup. Any base for the group G will be a base for the subgroup $H(P)$, so the aim is to construct a strong generating set relative to the base for G. That is, we need generators for $H(P)^{(k)}, H(P)^{(k-1)}, ..., H(P)^{(2)}, H(P)^{(1)}$.

The algorithms for enumerating all elements (or all base images) will enumerate the elements of $G^{(k)}$ first, then the elements of $G^{(k-1)} - G^{(k)}$, ..., $G^{(1)} - G^{(2)}$, provided that the identity element is the "first" element of the set of coset representatives $U^{(i)}$, for all i. Modifying these enumerations to be searches for elements satisfying the property P, will lead naturally to the construction of $H(P)^{(k)}, ..., H(P)^{(1)}$, and hence to a strong generating set of $H(P)$.

It is best to think of these searches as enumerations of all base images, because then the search tree can be pruned using points, sets of points, etc rather than using elements, sets of elements, etc. Points are simply easier to manipulate than elements.

Figure 1 shows the tree of base images $[\gamma_1, \gamma_2, ..., \gamma_k]$ for the symmetric group of degree 4. The enumeration traverses the tree in preorder. This is also called a depth-first or backtrack traversal, hence the term, *backtrack search.*

Figure 1 : Symmetric Group of Degree 4

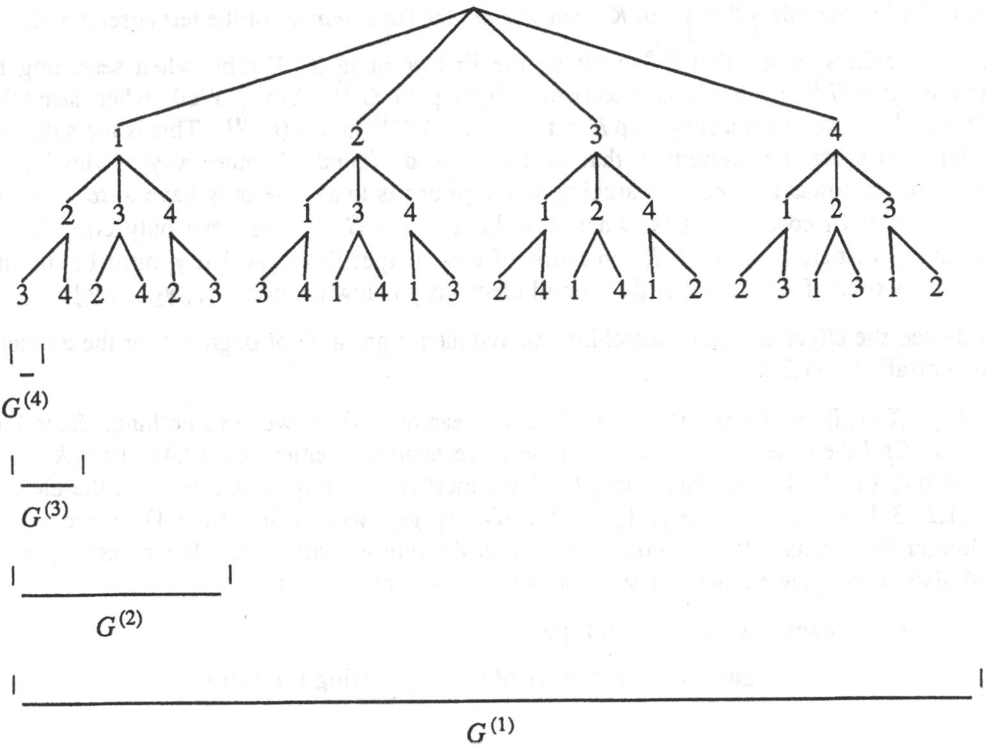

First in Orbit

As in the case of searching small groups, we can use cosets to prune the search of very large permutation groups. This is a problem-independent improvement. The information about cosets is translated into information about orbits, thus giving the first in the orbit criterion. To perform this translation, we must have a total order on the elements of the group.

Suppose we order the points of Ω, so that the base points $\beta_1, \beta_2, \ldots, \beta_k$ are the first k points of Ω. If we order the base images B^g in their lexicographical order, as follows,

$$B^g < B^{g'} \text{ if } [\beta_1^g, \beta_2^g, \ldots, \beta_i^g] = [\beta_1^{g'}, \beta_2^{g'}, \ldots, \beta_i^{g'}] \text{ and } \beta_{i+1}^g < \beta_{i+1}^{g'} \qquad (*)$$

and if we order the permutations g of G by their lexicographical order

$g < g'$ if g and g' have same images on the first i points of Ω
and the image of $i+1$st point under g
< the image of $i+1$st point under g'

then these two orders actually coincide. That is,

$$g < g' \text{ if and only if } B^g < B^{g'}.$$

Let K be a subgroup of G. The elements of the right coset $K \underset{K}{\times} g$ have base images $B^{K \times g}$ and the elements of the left coset $\underset{K}{g} \times K$ have base images $\left[B^g \right]$. In particular, if $\beta_1{}^g$ is *not* the first point in the orbit $\left[\beta_1{}^g \right]$ of K, then g is *not* the first element of the left coset $g \times K$.

This generalizes to say that $\beta_i{}^g$ must be the first point in its K-orbit when searching for elements $g \in G^{(i)}$ where K is a known subgroup of $G^{(i)}$. Why? Well, when searching $G^{(i)} - G^{(i+1)}$, we know a subgroup K between $H(P)^{(i+1)}$ and $H(P)^{(i)}$. This is the subgroup of elements satisfying property P that we have already found. Another way of viewing the discarding of cosets as used in searching small groups is to say we only have to test the *first* element of each coset. That is, when searching $G^{(i)} - G^{(i+1)}$ we need only consider the elements g that are first in $g \times K$. In terms of base images this says that γ_i must be the first point in its orbit of K (since γ_i *is* $\beta_i{}^g$, for all elements g below the node $[\gamma_1, \gamma_2, ..., \gamma_i]$).

Let us see the effect of this in searching the symmetric group G of degree 4 for the elements that centralize z=(1,3)(2,4).

Initially, $K=\{id\}$, so its orbits are irrelevant. Searching $G^{(3)}$ we find nothing. Searching $G^{(2)}$ we find the base image [1,4,3] and the corresponding element g_1=(2,4). Thus $K=\langle g_1 \rangle$ with orbits {1} {2,4} {3}. Searching $G^{(1)}$ we must consider $\gamma_1=2$, and we find the element g_2=(1,2)(3,4) with base image [2,1,4]. Thus $K=\langle g_1, g_2 \rangle$ with orbits {1,2,3,4}. Since $\gamma_1=2$ is no longer first in its orbit, we no longer search the subtree with $\gamma_1=2$. The choices $\gamma_1=3$ or $\gamma_1=4$ also do not give elements first in the coset, so we are finished.

The tree we have searched is given in Figure 2.

Figure 2 : Centralizer of (1,3)(2,4) using First in Orbit

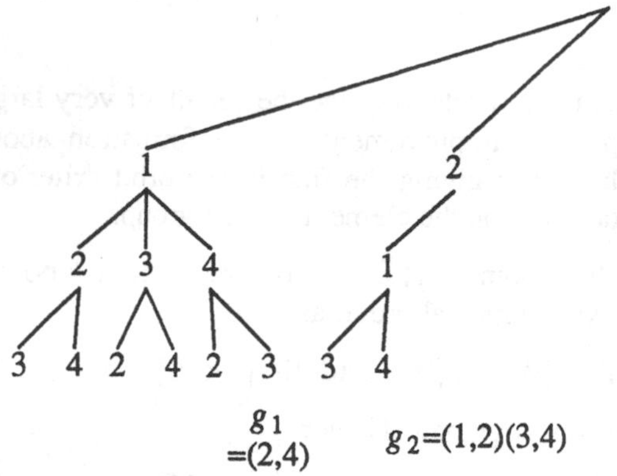

The tree we would have searched for the centralizer of $z=(1,2)(3,4)$ is given in Figure 3.

Figure 3 : Centralizer of (1,2)(3,4) using First in Orbit

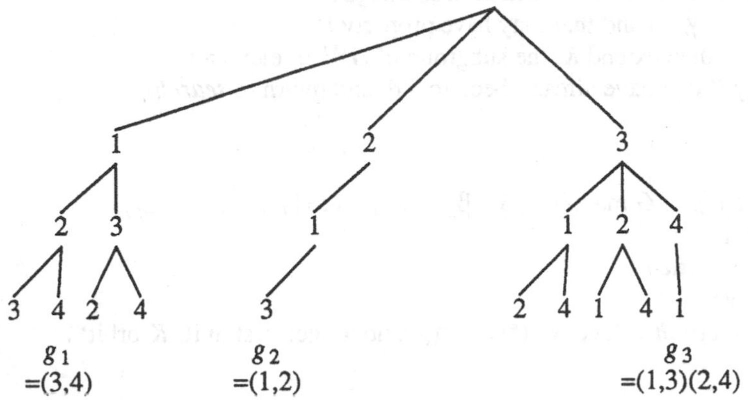

The search algorithm is presented in Algorithm 1 using two recursive procedures *search* and *generate*. In *generate* we wish to return through all the recursive levels of *generate* to procedure *search* once we have found a new element satisfying P. We do not strictly follow Pascal and include code to terminate each of the levels of recursion. Instead, we just indicate that this is what we wish to occur.

Algorithm 1 : Backtrack Search using First in Orbit

Input : a group G with a base $B=[\beta_1, \beta_2, ..., \beta_k]$ and a strong generating set S;
a decidable property P whose elements form a subgroup;

Output : a strong generating set relative to B for the subgroup $K = H(P)$;

procedure *search*(G : group; P : property; s : 1..k+1; **var** K : group);
(* Search $G^{(s)}$ for the subgroup K of elements satisfying P. *)
begin
  if $s = k+1$ **then**
    $K := \{id\}$;
  else
    search($G, P, s+1, K$);

    (*now search $G^{(s)} - G^{(s+1)}$ *)
    for each point $\gamma_s \in \Delta^{(s)}$ **do**
      if γ_s is first in its K-orbit **then**
        generate($G, P, s, s+1, [\beta_1, \beta_2, ..., \beta_{s-1}, \gamma_s], K$);
      end if;
    end for;
  end if;
end;

```
procedure generate( G : group; P : property; s : 1..k+1; i : 1..k+1;
                    [γ1, γ2, ..., γi-1] : initial segment of base image;
                    var K : group );
(*
  Generate the elements of G^(s) whose base image start
  with [γ1, γ2, ..., γi-1] and that may have property P.
  If one is found then extend K, the subgroup of G^(s) of elements
  with property P that have already been found, and return to search.
*)

begin
  find an element g ∈ G mapping [β1, β2, ..., βi-1] to [γ1, γ2, ..., γi-1];
  if i = k+1 then
    if g satisfies P then
      K := <K, g>;
      return to search at level s;  (*since γs is no longer first in its K-orbit*)
    end if;
  else
    for each point γi in [Δ^(i)]^g do
      generate( G, P, s, i+1, [γ1, γ2, ..., γi], K );
    end for;
  end if;
end;

begin
  search( G, P, 1, K );
end;
```

Restrictions on Image Points

This section looks at using the property P to prune the search. We are concerned with finding restrictions on the base images of elements that satisfy the property. Ideally, these restrictions should lead to a very restrictive test on the points γ_i of the base image, and the test should be quick and easy to compute.

We begin by looking at an example. Consider the property

$$P_z \ : \ g \in G \ \text{ centralizes } \ z \in G$$

defining the centralizer of z in G. Recall that a consequence of g centralizing z is that, for any point β,

$$\beta^{g \times z} \ = \ \beta^{z \times g}.$$

Suppose we apply this fact to the case where $\beta_i = \beta_j{}^z$. Then

$$\gamma_i = \beta_i{}^g \ = \ (\beta_j{}^g)^z \ = \ \gamma_j{}^z.$$

If $j < i$ then we know γ_j when we come to choose γ_i, so the choice for γ_i is restricted to at most one point, namely $\gamma_j{}^z$. Furthermore, suppose that $\beta_j{}^{z^r} = \beta_j$. Then $\gamma_j{}^{z^r} = \gamma_j$. Hence, the image γ_j of β_j is restricted to be a point in a z-cycle of the same length as the z-cycle containing β_j.

Figure 4 shows the result of applying this restriction to the construction of the centralizer of z=(1,3)(2,4) in the symmetric group of degree 4: the base is [1,2,3] so $\beta_3 = \beta_1{}^z$ which implies that $\gamma_3 = \gamma_1{}^z$. Figure 5 does the same for the centralizer of z=(1,2)(3,4), where $\beta_2 = \beta_1{}^z$.

Figure 4 : Centralizer of (1,3)(2,4) Restricting Choice of Image Points

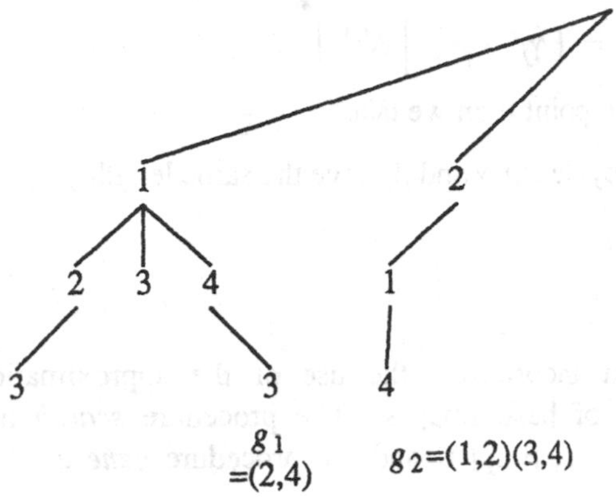

Figure 5 : Centralizer of (1,2)(3,4) Restricting Choice of Image Points

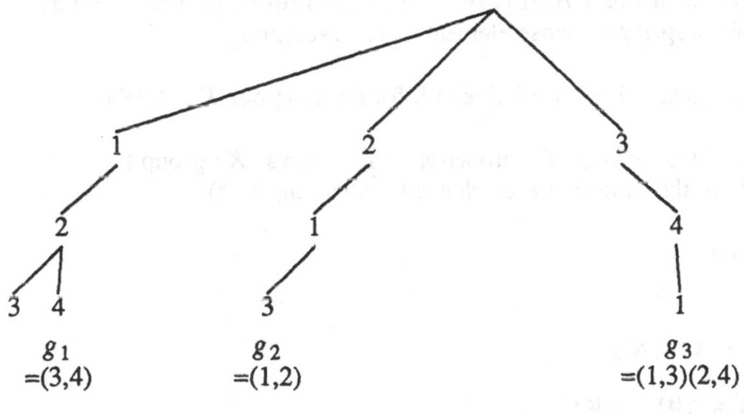

To incorporate such restrictions into the backtrack search we require information about the set

$$\Omega_P([\gamma_1, \gamma_2, ..., \gamma_{i-1}]) = \{ \gamma_i \mid \text{there is an element of } G \text{ satisfying property } P \text{ whose base image begins with } [\gamma_1, \gamma_2, ..., \gamma_i] \}.$$

In general, we will not know this set exactly. One fact we do know is that, for any element $g \in G$ mapping $[\beta_1, \beta_2, ..., \beta_{i-1}]$ to $[\gamma_1, \gamma_2, ..., \gamma_{i-1}]$,

$$\Omega_P([\gamma_1, \gamma_2, ..., \gamma_{i-1}]) \subseteq \left[\Delta^{(i)} \right]^g.$$

This is used as an upper bound on the set and its approximations. It will suffice for the backtrack search if we have an approximation

$$\overline{\Omega}_P([\gamma_1, \gamma_2, ..., \gamma_{i-1}]) = \{\ \gamma_i \mid \text{there may be an element of } G \text{ satisfying property } P \text{ whose base image begins with } [\gamma_1, \gamma_2, ..., \gamma_i]\ \}$$

which contains $\Omega_P([\gamma_1, \gamma_2, ..., \gamma_{i-1}])$. The approximation is chosen to be as restrictive as possible, while still remaining easily computable.

For the centralizer problem, where $\beta_i = \beta_j^{\,z}$, $j < i$, we take

$$\overline{\Omega}_{P_z}([\gamma_1, \gamma_2, ..., \gamma_{i-1}]) = \{\ \gamma_j^{\,z}\ \} \cap \left[\Delta^{(i)}\right]^g,$$

and if β_i is not the image of some previous base point then we take

$$\overline{\Omega}_{P_z}([\gamma_1, \gamma_2, ..., \gamma_{i-1}]) = \{\ \gamma \mid \text{the } z\text{-cycles of } \gamma \text{ and } \beta_i \text{ have the same length}\ \}$$

$$\cap \left[\Delta^{(i)}\right]^g.$$

Algorithm 2 presents the modifications that incorporate the use of the approximation $\overline{\Omega}_P([\gamma_1, \gamma_2, ..., \gamma_{i-1}])$ to restrict the choice of base images. The procedure *search* has replaced the reference to $\Delta^{(s)}$ by $\overline{\Omega}_P([\beta_1, \beta_2, ..., \beta_{i-1}])$; and the procedure *generate* has replaced the reference to $\left[\Delta^{(i)}\right]^g$ by $\overline{\Omega}_P([\gamma_1, \gamma_2, ..., \gamma_{i-1}])$.

Algorithm 2 : Backtrack Search Restricting Choice of Base Images

Input : a group G with a base $B=[\beta_1, \beta_2, ..., \beta_k]$ and a strong generating set S;
a decidable property P whose elements form a subgroup;

Output : a strong generating set relative to B for the subgroup $K = H(P)$;

```
procedure search( G : group; P : property; s : 1..k+1; var K : group );
(* Search G^(s) for the subgroup K of elements satisfying P. *)
begin
  if s = k+1 then
    K := {id};
  else
    search( G, P, s+1, K );

    (*now search G^(s) − G^(s+1) *)
    for each point γ_s ∈ Ω̄_P( β_1, β_2, ..., β_{s−1} ) do
      if γ_s is first in its K-orbit then
        generate( G, P, s, s+1, [β_1, β_2, ..., β_{s−1}, γ_s], K );
      end if;
    end for;
  end if;
end;
```

```
procedure generate( G : group; P : property; s : 1..k+1; i : 1..k+1;
                    [γ1, γ2, ..., γi-1] : initial segment of base image;
                    var K : group );
(* Generate the elements of G^(s) whose base image start
   with γ1, γ2, ..., γi-1 and that may have property P.
   If one is found then extend K, the subgroup of G^(s) of elements
   with property P that have already been found, and return to search. *)

begin
  find an element g ∈ G mapping [β1, β2, ..., βi-1] to [γ1, γ2, ..., γi-1];
  if i = k+1 then
    if g satisfies P then
      K := <K, g>;
      return to search at level s; (*since γs is no longer first in its K-orbit*)
    end if;
  else
    for each point γi in Ω̄P( [γ1, γ2, ..., γi-1] ) do
      generate( G, P, s, i+1, [γ1, γ2, ..., γi], K );
    end for;
  end if;
end;

begin
  search( G, P, 1, K );
end;
```

Choosing an Appropriate Base

By themselves, the restrictions on the images of base points are often not powerful enough. The most effective restrictions, like the centralizer restriction, use some relationship between the base points. However, the base that is given for the group may not display any of the required relationships between its points. So the development of effective backtrack searches involves an interplay between devising restrictions on the images of the base and devising ways of choosing a base that has the required relationships amongst its points in order for the restrictions to be effective.

Consider the example where G is the Mathieu group of degree 11. The group has a base [11,10,1,2]. For computing the centralizer of z=(2,5,3,9)(4,8,7,6), this base has no two base points β_i and β_j where $\beta_i = \beta_j{}^z$. Hence, the restrictions on base images will not be as effective as it might be. However, choosing the base [2,5,3,9] means that $\beta_i = \beta_{i-1}{}^z$, for i=2,3,4. This is the best we could have hoped for. Now choosing γ_1 uniquely determines the remainder of the base image of a centralizing element. The fact that the image of β_1=2 must lie in a z-cycle of length 4, and the first in the orbit test shows that the only values of γ_1 considered are 3 (leading to z^2 being found), and 4 (leading to (1,11)(2,4,9,6,3,7,5,8) being found).

Thus a group of order 7,920 has been searched by considering only 2 elements!

In general, the base appropriate to a problem is the one that minimises the size of the sets $\overline{\Omega}_P([\gamma_1, \gamma_2, ..., \gamma_{i-1}])$ that are considered. Therefore, the information needed to compute the sets is collected before the search commences, and is used to guide the selection of an appropriate base.

For the centralizer problem, we wish to have $\beta_i = \beta_{i-1}{}^z$ as often as possible, so we choose the first base points to be in a cycle of the longest length. If the points of that cycle do not form a base then we choose, from amongst the remaining cycles, one of the longest length, and so on, until we have a base.

Once we have selected a base, the base change algorithm, described in the next chapter, obtains a strong generating set relative to the chosen base.

Using a Known Subgroup

This section looks at using the information that is often provided free with the problem description. For example, in computing the centralizer of z in G, we have not used the fact that z is in the centralizer. The search finds z, or elements from which we can form z. Another example is the construction of normalizers. The normalizer $N_G(H)$ of H in G contains the subgroup H, so the elements of H do not have to be found by searching.

Let L be the subgroup of G we know to be in $H(P)$. The first thing we do when searching $G^{(s)} - G^{(s+1)}$ is to add $L^{(s)}$ to the subgroup K we are constructing. Generally, we will know a strong generating set T of L, so this amounts to including the elements of $T^{(s)} - T^{(s+1)}$ in the strong generators of K. If we do not know a strong generating set of L then we just include whatever elements of $L^{(s)}$ we can easily lay our hands on. For example, the generators of L which fix the first $s-1$ base points.

The modifications to the backtrack search only affect the procedure *search*. These changes are presented in Algorithm 3.

Algorithm 3 : Backtrack Search Using Known Subgroup

Input : a group G with a base B=[β_1, β_2, ..., β_k] and a strong generating set S;
a decidable property P whose elements form a subgroup;
a subgroup L of G whose elements satisfy P;
Output : a strong generating set relative to B for the subgroup $K = H(P)$;

procedure *search*(G : group; P : property; s : 1..k+1; **var** K : group);
(* Search $G^{(s)}$ for the subgroup K of elements satisfying P. *)
begin
  if $s = k+1$ **then**
    $K := \{id\}$;
  else
    search($G, P, s+1, K$);
    add $L^{(s)}$ to K; (*use known subgroup in search of $G^{(s)} - G^{(s+1)}$ *)
    for each point $\gamma_s \in \overline{\Omega}_P([\beta_1, \beta_2, ..., \beta_{s-1}])$ **do**
      if γ_s **is first in its** K**-orbit then**
        generate($G, P, s, s+1, [\beta_1, \beta_2, ..., \beta_{s-1}, \gamma_s], K$);
      end if;
    end for;
  end if;
end;

Searching Images of an Initial Base Segment

A special case of a known subgroup is a stabiliser $G^{(l)}$ of the stabiliser chain. This leads to even more improvement than given in the last section because now we do not need to generate a complete base image but instead only the images of the first $l-1$ base points. The reason for this is that each such image determines a coset of $G^{(l)}$. We have $G^{(l)}$ in the subgroup K, so we need only test one element from each coset. There is no need to generate all the elements of that coset by generating the complete base images which extend the image of the initial base segment.

An example of this phenomenom is constructing the centralizer of z=(1,2) in the symmetric group of degree 4. The permutations which fix 1 and 2 must centralize z, so $G^{(3)}$ is a known subgroup of $H(P)$.

Another common example is the construction of set stabilisers. Here the group that stabilises all the points in the set individually is arranged to be a member of the stabiliser chain. It is a known subgroup of $H(P)$, and only the images of the points in the set have to be generated rather than the complete base image.

The modifications to Algorithm 2 are presented in Algorithm 4.

Algorithm 4 : Backtrack Search Using Shortened Base Images

Input : a group G with a base $B=[\beta_1, \beta_2, ..., \beta_k]$ and a strong generating set S;
a decidable property P whose elements form a subgroup;
Output : a strong generating set relative to B for the subgroup $K = H(P)$;

```
procedure search( G : group; P : property; s : 1..k+1; l : 1..k+1; var K : group );
(* Search G^(s) for the subgroup K of elements satisfying P.
   It is known that the elements of G^(l) satisfy P. *)
begin
  if s = l then
    K := G^(l);
  else
    search( G, P, s+1, l, K );
    (*now search G^(s) - G^(s+1) *)
    for each point γ_s ∈ Ω̄_P( [β_1, β_2, ..., β_{s-1}] ) do
      if γ_s is first in its K-orbit then
        generate( G, P, s, l, s+1, [β_1, β_2, ..., β_{s-1}, γ_s], K );
      end if;
    end for;
  end if;
end;

procedure generate( G : group; P : property; s : 1..k+1; l : 1..k+1; i : 1..k+1;
                    [γ_1, γ_2, ..., γ_{i-1}] : initial segment of base image;
                    var K : group );
(* Generate the coset representatives of G^(l) in G^(s) whose base image start
   with [γ_1, γ_2, ..., γ_{i-1}] and that may have property P.
   If one is found then extend K, the subgroup of G^(s) of elements
   with property P that have already been found, and return to search. *)
begin
  find an element g ∈ G mapping [β_1, β_2, ..., β_{i-1}] to [γ_1, γ_2, ..., γ_{i-1}];
  if i = l then
    if g satisfies P then
      K := <K, g>;
      return to search at level s;  (*since γ_s is no longer first in its K-orbit*)
    end if;
  else
    for each point γ_i in Ω̄_P( [γ_1, γ_2, ..., γ_{i-1}] ) do
      generate( G, P, s, l, i+1, [γ_1, γ_2, ..., γ_i], K );
    end for;
  end if;
end;
```

begin
choose an appropriate base for the problem;
let $G^{(l)}$ be the largest stabiliser whose elements have property P;
search($G, P, 1, l, K$);
end;

More on Cosets

This section continues the discussion on using cosets to prune the search. This problem-independent heuristic has already lead to the powerful first in the orbit test. There is still more to be gained, however, even though these improvements are not as significant as the first in the orbit test.

The equivalence of the first element g in the left coset of K and the first base image in $B^{g \times K}$ easily leads to

Proposition 1

An element g with base image $[\gamma_1, \gamma_2, ..., \gamma_k]$ is first in the left coset $g \times K$ if and only if, for each i, γ_i is the first point in the orbit of γ_i under $K_{\gamma_1, \gamma_2, ..., \gamma_{i-1}}$.

Corollary

If γ_i is not the first point in its $K_{\gamma_1, \gamma_2, ..., \gamma_{i-1}}$-orbit then no element g whose base image begins $[\gamma_1, \gamma_2, ..., \gamma_i]$ is first in the left coset $g \times K$.

Corollary

If $K \le G^{(s)}$ and $g \in G^{(s)}$, and g maps β_s to a point γ_s which is not first in its K-orbit, then g is not first in the left coset $g \times K$.

The base change algorithm does allow us to compute $K_{\gamma_1, \gamma_2, ..., \gamma_{i-1}}$. However, the cost is not trivial. For this reason, almost all backtracks do not use the complete test for an element being first in its left coset of K. One exception is the algorithm for computing the automorphism group of a group. The conclusions of the first corollary also hold when we know some subgroup K' of the stabiliser $K_{\gamma_1, \gamma_2, ..., \gamma_{i-1}}$. A typical example is the subgroup generated by those generators of K which fix $\gamma_1, \gamma_2, ..., \gamma_{i-1}$.

The equivalence of the first element in the right coset and the first image in $B^{K \times g}$ easily leads to

Proposition 2

An element g with base image $[\gamma_1, \gamma_2, ..., \gamma_k]$ is the first element in the right coset $K \times g$ if and only if, for each i, and for each $\gamma \in \beta_i^{K^{(i)}}$, $\gamma_i \le \gamma^g$.

Corollary

If $\beta_i \in \beta_j^{K^{(j)}}$, for $j < i$, and $\gamma_i < \gamma_j$ then any element g whose base image begins with $[\gamma_1, \gamma_2, ..., \gamma_i]$ is not the first element in the right coset $K \times g$.

In this instance, we know $K^{(i)} = K_{\beta_1, \beta_2, ..., \beta_{i-1}}$ because the backtrack produces a strong generating set of K relative to $B = [\beta_1, \beta_2, ..., \beta_k]$. However, the test requires that later base points β_i be in the basic orbit $\beta_j^{K^{(j)}}$ of K. It is unclear how frequently this occurs.

The only changes are to the procedure *generate*. They are given in Algorithm 5.

Algorithm 5 : Backtrack Search Using Left and Right Cosets

Input : a group G with a base B=[$\beta_1, \beta_2, ..., \beta_k$] and a strong generating set S;
  a decidable property P whose elements form a subgroup;

Output : a strong generating set relative to B for the subgroup $K = H(P)$;

procedure *generate*(G : group; P : property; s : 1..k+1; l : 1..k+1; i : 1..k+1;
    [$\gamma_1, \gamma_2, ..., \gamma_{i-1}$] : initial segment of base image;
    var K : group);
(* Generate the coset representatives of $G^{(l)}$ in $G^{(s)}$ whose base image start
  with [$\gamma_1, \gamma_2, ..., \gamma_{i-1}$] and that may have property P.
  If one is found then extend K, the subgroup of $G^{(s)}$ of elements
  with property P that have already been found, and return to *search*. *)
begin
  find an element $g \in G$ mapping [$\beta_1, \beta_2, ..., \beta_{i-1}$] to [$\gamma_1, \gamma_2, ..., \gamma_{i-1}$];
  if $i = l$ **then**
    if g satisfies P **then**
      $K := \langle K, g\rangle$;
      return to *search* at level s; (*since γ_s is no longer first in its K-orbit*)
    end if;
  else
    for each point γ_i in $\overline{\Omega}_P$([$\gamma_1, \gamma_2, ..., \gamma_{i-1}$]) **do**
      if (γ_i is first in $\gamma_i^{K_{\gamma_1, \gamma_2, ..., \gamma_{i-1}}}$)
          and
        ($\gamma_i \geq \gamma_j$, for all $j < i$ where $\beta_i \in \beta_j^{K^{(j)}}$) **then**
        generate(G, P, s, l, i+1, [$\gamma_1, \gamma_2, ..., \gamma_i$], K);
      end if;
    end for;
  end if;
end;

Preprocessing

There may be many calculations of the approximations $\overline{\Omega}_P$ performed during the backtrack search. Typically, these computations will use the same information. To speed up the computation, this information should be readily available rather than recalculated each time. Also, as we have mentioned earlier, such information is used in the selection of an appropriate base, so its computation at the outset is necessary in any case.

An example, which occurs in the construction of centralizers, are the points in a z-cycle of the same length of the z-cycle of β_i, when $\beta_i \neq \beta_j^g$, for some $j < i$. This set is constant throughout the computation, so it may be calculated at the outset and stored for later repeated use in calculating $\overline{\Omega}_P$([$\gamma_1, \gamma_2, ..., \gamma_{i-1}$]).

Case Study 1 : Centralizer

The centralizer has been used as our running example, so this section will just be a summary. The property is

$$P_z \ : \ g \in G \ \text{centralizes} \ z \in G$$

The base is chosen to match the cycles of z, with the longest cycles being matched first. The restrictions on the base images are

$$\overline{\Omega}_{P_z}([\gamma_1, \gamma_2, ..., \gamma_{i-1}]) = \{ \gamma_{i-1}{}^z \} \cap \left[\Delta^{(i)} \right]^g,$$

if $\beta_i = \beta_{i-1}{}^z$, and

$$\overline{\Omega}_{P_z}([\gamma_1, \gamma_2, ..., \gamma_{i-1}]) = \{ \gamma \mid \text{the } z\text{-cycles of } \gamma \text{ and } \beta_i \text{ have the same length} \}$$

$$\cap \left[\Delta^{(i)} \right]^g$$

otherwise.

The known subgroup is $<z>$.

Shortened base images are used if the process of matching the cycles of z eventually considers cycles of length one. In this case, the end of the base will consist of fixed points of z. The stabiliser of the fixed points centralizes z, so we only generate the images of the initial segment of the base.

Preprocessing constructs, for the relevant values of i, the set of points in z-cycles of length $|\beta_i{}^{<z>}|$,

Case Study 2 : Normalizer

This section considers the normalizer of a subgroup H of G. We assume we have a base and strong generating set for H. The property is

$$P_H \ : \ g \in G \ \text{normalizes} \ H \leq G$$

The two facts we know about normalizing elements which lead to our choice of a base and the restrictions on the base images are

Lemma 1

If g normalizes a subgroup H then g permutes the orbits of H.

Lemma 2

If g normalizes a subgroup H and g fixes β then g normalizes the stabiliser H_β.

The lemmas imply that, when searching $G^{(s)}$, the elements of the normalizer will normalize each of $H^{(s)}, H^{(s-1)}, ..., H^{(1)}$ and permute the orbits of each of the subgroups. This means that

(i) the length of the orbit of γ_i is the same as the length of the orbit of β_i, and

(ii) β_i and β_j are in the same orbit if and only if γ_i and γ_j are in the same orbit,

for each subgroup $H^{(s)}, H^{(s-1)}, ..., H^{(1)}$. So we define

$$\overline{\Omega}_{P_H}([\gamma_1, \gamma_2, ..., \gamma_{i-1}]) = \left[\Delta^{(i)}\right]^g$$

$$\cap \bigcap_{t=1}^{s} \{ \gamma \in \Omega \mid |\gamma^{H^{(t)}}| = |\beta_i^{H^{(t)}}| \}$$

$$\cap \bigcap_{j=1}^{i-1} \{ \gamma_j^{H^{(t)}} \mid 1 \le t \le s \text{ and } \beta_i \in \beta_j^{H^{(t)}} \}$$

$$- \bigcup_{j=1}^{i-1} \{ \gamma_j^{H^{(t)}} \mid 1 \le t \le s \text{ and } \beta_i \notin \beta_j^{H^{(t)}} \}$$

where, as usual, g is an element of G mapping $[\beta_1, \beta_2, ..., \beta_{i-1}]$ to $[\gamma_1, \gamma_2, ..., \gamma_{i-1}]$. In support of these restrictions on the base images, the base is chosen with three objectives in mind :

(1) maximise the number of occurrences of $\beta_i \in \beta_{i-1}{}^{H^{(i-1)}}$,

(2) minimise the number of $H^{(i)}$-orbits of length $|\beta_i{}^{H^{(i)}}|$, and

(3) minimise the number of points in the $H^{(i)}$-orbits of length $|\beta_i{}^{H^{(i)}}|$.

The known subgroup is H.

If the process of choosing a base eventually includes fixed points of H at the end of the base, then only images of an initial segment of the base are generated, since the stabiliser of these fixed points normalizes H. Indeed, the stabiliser *centralizes* H.

The preprocessing stores information about the orbits of $H, H^{(2)}, H^{(3)}, ...$ such as

(a) the length, and a representative point of each orbit,

(b) each orbit as a linked list of points,

(c) the fusion of the orbits of $H^{(t)}$ in $H^{(t-1)}$, for $t > 1$,

(d) which orbits contain the base points, and their lengths, and

(e) which pairs of base points are contained in the same orbit.

Case Study 3 : Intersection

This section considers the intersection of two groups $H1$ and $H2$ of permutations on Ω. We assume we have a base and strong generating set for both groups. The property is

$$P_{H1 \cap H2} : g \in H1 \text{ and } g \in H2.$$

For either group $H1$ or $H2$ we could generate all the base images, and therefore all the elements, and test if the element was in the other group. However, if we arrange for both groups to have the same base, we can in effect compare base images directly. A base image will then correspond to an element in the intersection only if it can be generated in both $H1$ and $H2$. The search generates both groups simultaneously, or at least enough of both groups to determine the intersection. The restrictions we place on the base images are

$$\overline{\Omega}_{P_{H1 \cap H2}}([\gamma_1, \gamma_2, ..., \gamma_{i-1}]) = \left[\Delta 1^{(i)}\right]^{g1} \cap \left[\Delta 2^{(i)}\right]^{g2}$$

where $g1$ (respectively $g2$) is an element of $H1$ (respectively $H2$) mapping the first $i-1$ base points to $[\gamma_1, \gamma_2, ..., \gamma_{i-1}]$. (The numbers 1 and 2 also distinguish the basic orbits of the two groups.)

Once the complete base image is generated, both $g1$ and $g2$ are uniquely determined. However, we still must see if they are the same permutation before concluding that the base image gives an element in the intersection.

Case Study 4 : Set Stabiliser

This section considers the stabiliser of a set $\Delta = \{\delta_1, \delta_2, ..., \delta_{l-1}\}$ of points. The property is

$$P_\Delta : g \in G \text{ and } \delta^g \in \Delta, \text{ for all } \delta \in \Delta.$$

The property places restrictions on the images of the points in the set, so the base is chosen to begin with $\delta_1, \delta_2, ..., \delta_{l-1}$ and the approximation is

$$\overline{\Omega}_{P_\Delta}([\gamma_1, \gamma_2, ..., \gamma_{i-1}]) = \Delta \cap \left[\Delta^{(i)}\right]^{g}.$$

The stabiliser $G^{(l)}$ fixes each point of the set Δ, so it is in the set stabiliser. Therefore, only images of the initial segment of the base are generated.

Summary

The problem-independent heuristics for improving the backtrack search rely on the use of cosets. They are

(1) first in the K-orbit test in *search*,

(2) first in the $K_{\gamma_1, \gamma_2, ..., \gamma_{i-1}}$-orbit test in *generate*, and

(3) weak test for first in the right coset of K.

Of these, the first is of great significance, leading to great improvements in the backtrack search.

The problem-dependent heuristics rely on the property P for their details, though their general strategies are applicable to a wide class of problems. They are

(1) restricting the choice of images using $\overline{\Omega}_P$,

(2) choosing a base appropriate to the problem,

(3) using any known subgroups, and

(4) generating images of an initial base segment.

There is a close connection between the first two. Indeed, the effectiveness of (1) depends on (2), and the selection of a base depends on how it will be used in (1). Both offer significant improvements in the backtrack searches that have been discussed. The importance of (3) and (4) depends on the problem.

The algorithms for the construction of centralizers, normalizers, intersections, and set stabilisers have been presented. The techniques of this chapter are also used in the construction of the automorphism group of a group.

Exercises

(1/Moderate) This series of exercises deals with backtrack searches within the symmetries of the projective plane of order two. They have been chosen so that [1,2,4] is an appropriate base, though maybe not the most appropriate. Therefore, no base changes are necessary. In each case, draw the search tree. Choose an appropriate algorithm that uses at least the first in the K-orbit test, the restriction on images, and any known subgroup.

(i) Let $z = (1,2)(4,7)$. Determine the centralizer of z.

(ii) Let $H = \langle (1,2)(4,7), (4,7)(5,6) \rangle$ of order 4. Determine its normalizer.

(iii) Let $H1 = \langle (1,2,4,7)(3,6), (2,4,7)(3,5,6), (4,7)(5,6) \rangle$ and let $H2 = \langle (1,5,7,3)(2,4), (2,3)(4,5), (2,5)(3,4) \rangle$ both of order 24. Determine their intersection.

(iv) Show that the normalizer of $H1$ is $H1$, and the normalizer of $H2$ is $H2$.

(v) Determine the stabiliser of the set $\{1,2,4\}$.

Bibliographical Remarks

Backtrack searches developed as part of the folklore of combinatorial computing. The first reference to the term *backtrack* in this context appears to be R. J. Walker, "*An enumerative technique for a class of combinatorial problems*", **Combinatorial Analysis**, R. E. Bellman and M. Hall, Jr (editors), Proceedings of the Symposium on Applied Mathematics, **10**, 1960, 91-94.

The use of backtrack searches in relation to permutation groups began with the development of the centralizer algorithm in C. C. Sims, "*Determining the conjugacy classes of a permutation group*", **Computers in Algebra and Number Theory** (Proceedings of the Symposium on Applied Mathematics, New York, 1970), G. Birkhoff and M. Hall, Jr (editors), SIAM-AMS Proceedings, volume 4, American Mathematics Society, Providence, Rhode Island, 1971, 191-195. In this paper, Sims places restrictions on the images of the base, chooses an appropriate base, uses the first in the *K*-orbit test, and uses the weak test for first in the right coset. The classification of the first element in the right coset in terms of its base image is also due to this paper. The next paper developed the intersection algorithm : C. C. Sims, "*Computation with permutation groups*", (Proceedings of the Second Symposium on Symbolic and Algebraic Manipulation, Los Angeles, 1971), S. R. Petrick (editor), Association of Computing Machinery, New York, 1971, 23-28, and the ideas behind the set stabiliser algorithm are found in the lectures Sims gave in Oxford in January and February 1973.

The classification of the first element of the left coset in terms of its base image can be found in G. Butler, "*Computing in permutation and matrix groups II : backtrack algorithm*", Mathematics of Computation **39**, 160 (1982) 671-670. This chapter follows the presentation of this paper, but recasts the algorithms in a modern language and uses recursion. Although this chapter has concentrated on constructing subgroups, the backtrack search can also be used to find single elements with a given property, as discussed in the above paper.

The normalizer algorithm is due to G. Butler, "*Computing normalizers in permutation groups*", Journal of Algorithms **4** (1983) 163-175, and an algorithm for computing the automorphism group of a group using a backtrack search is described in H. Robertz, **Eine Methode zur Berechnung der Automorphismengruppe einer endliche Gruppe**, Diplomarbeit, Aachen, 1976.

Little is known about the analysis of backtrack algorithms. M. Fontet, "*Calcul de centralisateur d'un groupe de permutations*", Bulletin de la Société Mathématique de France Mémoire **49-50** (1977) 53-63, shows that computing the centralizer of an element in the symmetric group is polynomial-time. The intersection algorithm is known to be exponential by C. M. Hoffman, "*On the complexity of intersecting permutation groups and its relationship with graph isomorphism*", Technical Report 4/80, Institut für Informatik und Praktische Mathematik, Christian-Albrechts-Universität Kiel, 1980. Hoffman presents polynomial algorithms for special cases of the intersection problem. Estimating the running time of a particular backtrack algorithm on a given problem is discussed in D. E. Knuth, "*Estimating the efficiency of backtrack programs*", Mathematics of Computation **29** (1975) 121-136.

A backtrack search, similar to the one described here, is also used to compute automorphism groups of combinatorial objects such as graphs, codes, and Hadamard matrices. Some references are B. D. McKay, "*Computing automorphisms and canonical labelling of graphs*", **Combinatorial Mathematics,** (Proceedings of the International Conference on Combinatorial Theory, Canberra, August 16-27, 1977), D. A. Holton and J. Seberry (editors), Lecture Notes in Mathematics, **686**, Springer-Verlag, Berlin, 1978, 223-232; J. S. Leon, "*An algorithm for computing the automorphism group of a Hadamard matrix*", Journal of Combinatorial Theory, Series A, **27** (1979) 289-306; J. S. Leon, "*Computing automorphism groups of error-correcting codes*", IEEE Transactions on Information Theory, **IT-28** (1982) 496-511; and J. S. Leon, "*Computing automorphism groups of combinatorial objects*", **Computational Group Theory,** (Proceedings of LMS Symposium on Computational Group Theory, Durham, 1982), M. Atkinson (editor), Academic Press, New York, 1984, 321-335. The last reference is a general overview of backtrack techniques for constructing automorphism groups of combinatorial objects.

Chapter 12. Base Change

This chapter answers the one difficulty arising from backtrack searches : Suppose we have selected an appropriate base, how do we obtain a strong generating set relative to the selected base? We assume we have some base and a strong generating set relative to that base, so we are not in the situation of starting just with the generators of the group. (That situation is discussed in the next chapter.)

Other Bases

Suppose $B = [\beta_1, \beta_2, ..., \beta_k]$ is a base for G, and that S is a strong generating set of G relative to B. What other bases can we easily obtain, and are strong generating sets relative to these other bases also easy to obtain? We present some examples. They are all rather obvious, except the last one.

(1) We can add points to the end of B.

For any point β_{k+1}, not in B, the sequence $\overline{B} = [\beta_1, \beta_2, ..., \beta_k, \beta_{k+1}]$ is a base for G, and S is a strong generating set of G relative to $\overline{B}$.

(2) We can delete redundant base points from the base B.

If $G^{(i)} = G^{(i+1)}$ then β_i is a redundant base point. So $\overline{B} = [\beta_1, \beta_2, ..., \beta_{i-1}, \beta_{i+1}, \beta_{i+2}, ..., \beta_k]$ is a base for G, and S is a strong generating set of G relative to $\overline{B}$.

(2′) We can delete redundant base points from the end of the base B.

If $G^{(k)} = \{id\}$, then $\overline{B} = [\beta_1, \beta_2, ..., \beta_{k-1}]$ is a base for G and S is a strong generating set of G relative to $\overline{B}$.

(3) We can interchange adjacent base points.

The sequence $\overline{B} = [\beta_1, \beta_2, ..., \beta_{j-1}, \beta_{j+1}, \beta_j, \beta_{j+2}, \beta_{j+3}, ..., \beta_k]$ obtained by interchanging the base points β_j and β_{j+1} is a base for G. A strong generating set of G relative to this base is not obtained trivially. A method for constructing a strong generating set is given in a later section.

(4) We can take the image of a base.

If $g \in G$ then the base image $\overline{B} = [\beta_1^g, \beta_2^g, ..., \beta_k^g]$ is a base for G. Furthermore, the conjugate S^g of S is a strong generating set of G relative to the base $\overline{B}$.

To show that the image of the base is a base, and that the conjugate of the strong generating set is the required strong generating set relative to the base image, we show that

$$G_{\beta^g} = g^{-1} \times \left[G_\beta \right] \times g.$$

If h fixes β^g then $g \times h \times g^{-1}$ fixes β, since

$$\beta \xrightarrow{\ g\ } \beta^g \xrightarrow{\ h\ } \beta^g \xrightarrow{\ g^{-1}\ } \beta$$

Hence, $h \in \left[G_\beta \right]^g$. Conversely, if $h \in G_\beta$ then $g^{-1} \times h \times g$ fixes β^g. Hence,

$$G_{\beta^g} = g^{-1} \times \left[G_\beta \right] \times g.$$

Thus, $G_{\beta_1^g, \beta_2^g, \ldots, \beta_k^g}$ is the conjugate of $G_{\beta_1, \beta_2, \ldots, \beta_k}$, which is the identity. Hence, the base image is a base. Furthermore, if the set T generates G_β then the set T^g generates $\left[G_\beta \right]^g$, which is G_{β^g}. Therefore, the conjugate of S contains generating sets for each group in the new stabiliser chain. That is, S^g is a strong generating set relative to $\overline{B}$.

Using (1), (2′), and (3) we can obtain any base, and a strong generating set relative to that base. Using (1), the points of the new base may be appended to the original base. Using (3) repeatedly, these new base points are moved into their correct position, and (2′) deletes the now redundant points of the original base from the end.

We will see later that this approach can be improved through the use of (4).

Interchanging Adjacent Base Points

This section shows how to construct a strong generating set relative to a base obtained by interchanging adjacent base points.

Suppose we are interchanging the base points β_j and β_{j+1}. To make our notation clear, all objects relative to the original base will be as per usual, while those relative to the new base $\overline{B} = [\beta_1, \beta_2, \ldots, \beta_{j-1}, \beta_{j+1}, \beta_j, \beta_{j+2}, \beta_{j+3}, \ldots, \beta_k]$ will have a bar $^-$. Thus,

$$\overline{G}^{(j+1)} = G_{\overline{\beta}_1, \overline{\beta}_2, \ldots, \overline{\beta}_j} = G_{\beta_1, \beta_2, \ldots, \beta_{j-1}, \beta_{j+1}} = G^{(j)}_{\beta_{j+1}}.$$

In fact, $\overline{G}^{(j+1)}$ is the only stabiliser in the new stabiliser chain that is different from its counterpart in the old stabiliser chain. That is, $\overline{G}^{(i)} = G^{(i)}$, for all $i \neq j+1$. The only basic orbits that are different are

$$\overline{\Delta}^{(j)} = \overline{\beta}_j^{\,\overline{G}^{(j)}} = \beta_{j+1}^{\,G^{(j)}}$$

and

$$\overline{\Delta}^{(j+1)} = \overline{\beta}_{j+1}^{\,\overline{G}^{(j+1)}} = \beta_j^{\,\overline{G}^{(j+1)}}.$$

The strong generating set S contains generators for all the stabilisers $\overline{G}^{(i)}$, except perhaps $\overline{G}^{(j+1)}$. Our aim is to construct a set of generators T of $\overline{G}^{(j+1)}$. Then $S \cup T$ is a strong generating set relative to the new base $\overline{B}$.

We know that

$$G^{(j+2)} \leq \overline{G}^{(j+1)} \leq G^{(j)}{}_{\beta_{j+1}} \leq G^{(j)}.$$

An element g of $G^{(j)}$ is in $\overline{G}^{(j+1)}$ if it maps β_j to some point in the basic orbit $\Delta^{(j)}$ and fixes β_{j+1}. We do not care what it does to the remaining base points $\beta_{j+2}, \beta_{j+3}, \ldots, \beta_k$. That is,

$$\beta_j \xrightarrow{\quad g \quad} \gamma \in \Delta^{(j)}$$

$$\beta_{j+1} \xrightarrow{\qquad} \beta_{j+1}$$

$$\cdots$$

One way to think of the problem is as a miniature search : Find elements of G whose base images start with $[\beta_1, \beta_2, ..., \beta_{j-1}, *, \beta_{j+1}]$, where "*" matches any point. A simple algorithm for this based on previous algorithms for searching and for enumerating all the elements of the group is presented as Algorithm 1. The algorithm does not return a set of generators of $\overline{G}^{(j+1)}$, it returns a set of coset representatives of $G^{(j+2)}$ in $\overline{G}^{(j+1)}$. The next algorithm shows it is a simple matter to form a set of generators using the coset representatives.

Algorithm 1 : Miniature Search

Input : a group G;
  a base $[\beta_1, \beta_2, ..., \beta_k]$ for G and a strong generating set;
  an integer j between 1 and $k-1$;

Output : a set T of coset representatives of $G^{(j+2)}$ in $\overline{G}^{(j+1)}$;

begin

T := empty set;

for each $\gamma \in \Delta^{(j)}$ **do**

find $g_1 \in G^{(j)}$ mapping β_j to γ;

if $\beta_{j+1} \in \left[\Delta^{(j+1)}\right]^{g_1}$ **then**
  choose $g_2 \in G^{(j+1)}$ mapping β_{j+1} to $\beta_{j+1}^{g_1^{-1}}$;
  add $g_2 \times g_1$ to T;
end if;

end for;

end;

Of course, choosing $\gamma = \beta_j$ leads to an element of $G^{(j+2)}$ being added to T, and choosing $\gamma = \beta_{j+1}$ does not lead to a permutation. To obtain a generating set for $\overline{G}^{(j+1)}$ we require generators for $G^{(j+2)}$ as well as the coset representatives of $G^{(j+2)}$ in $\overline{G}^{(j+1)}$. We know that $S^{(j+2)}$ generates $G^{(j+2)}$. Furthermore, as in earlier backtrack searches, we need only consider the points γ that are first in the orbit. In this case, the orbit under $<T>$. Combining these ideas leads to Algorithm 2.

Algorithm 2 : Search for Generators

Input : a group G;
  a base $[\beta_1, \beta_2, ..., \beta_k]$ for G and a strong generating set;
  an integer j between 1 and $k-1$;

Output : a set T of generators of $\overline{G}^{(j+1)}$;

begin

$T := S^{(j+2)}$; $\Gamma := \Delta^{(j)} - \{\beta_j, \beta_{j+1}\}$;

while $\Gamma \neq$ empty **do**

choose $\gamma \in \Gamma$; find $g_1 \in G^{(j)}$ mapping β_j to γ;
if $\beta_{j+1}^{g_1^{-1}} \in \Delta^{(j+1)}$ **then**
  find $g_2 \in G^{(j+1)}$ mapping β_{j+1} to $\beta_{j+1}^{g_1^{-1}}$;
  add $g_2 \times g_1$ to T; $\Gamma := \Gamma - \gamma^{<T>}$;
else
  $\Gamma := \Gamma - \gamma^{<T>}$;
end if;

end while;

end;

The next improvement comes about because we can determine when the search is finished. We know the order of the subgroup $\overline{G}^{(j+1)}$ that we are searching for because the equations

$$|G| = \prod_{i=1}^{k} |\Delta^{(i)}| = \prod_{i=1}^{k} |\overline{\Delta}^{(i)}|$$

show that

$$|\overline{\Delta}^{(j+1)}| = \frac{|\Delta^{(j)}| \times |\Delta^{(j+1)}|}{|\beta_{j+1}^{G^{(j)}}|}.$$

Hence, the search can be terminated once sufficient generators have been added to T to give an orbit of the correct size. The final algorithm is Algorithm 3. We have presented it as a procedure *interchange* for use by the complete base change algorithm.

Algorithm 3 : Interchange Base Points

Input : a group G;
a base $[\beta_1, \beta_2, ..., \beta_k]$ for G and a strong generating set;
an integer j between 1 and $k-1$;

Output : a base $\bar{B} = [\beta_1, \beta_2, ..., \beta_{j-1}, \beta_{j+1}, \beta_j, \beta_{j+2}, \beta_{j+3}, ..., \beta_k]$ for G;
a strong generating set relative to $\bar{B}$;

procedure *interchange*(**var** B : sequence of points;
var S : set of elements;
j : $1..k-1$);
(* Given a base $B = [\beta_1, \beta_2, ..., \beta_k]$ and a strong generating set S of G,
return the base obtained by interchanging β_j and β_{j+1},
and return a strong generating set relative to the new base. *)
begin
(*find generators T for $\bar{G}^{(j+1)}$ *)
compute $|\bar{\Delta}^{(j+1)}|$; $T := S^{(j+2)}$; $\Gamma := \Delta^{(j)} - \{\beta_j, \beta_{j+1}\}$;
$\Delta := \{\beta_j\}$; (* $= \beta_j^{<T>}$ and will grow to be $\bar{\Delta}^{(j+1)}$ *)

while $|\Delta| \neq |\bar{\Delta}^{(j+1)}|$ **do**

choose $\gamma \in \Gamma$; find $g_1 \in G^{(j)}$ mapping β_j to γ;
if $\beta_{j+1}^{g_1^{-1}} \in \Delta^{(j+1)}$ **then**
find $g_2 \in G^{(j+1)}$ mapping β_{j+1} to $\beta_{j+1}^{g_1^{-1}}$;
add $g_2 \times g_1$ to T; $\Delta := \beta_j^{<T>}$; $\Gamma := \Gamma - \Delta$;
else
$\Gamma := \Gamma - \gamma^{<T>}$;
end if;

end while;

(*return new base and strong generating set*)
(*only orbits and Schreier vectors at levels j and $j+1$ change*)
$B := [\beta_1, \beta_2, ..., \beta_{j-1}, \beta_{j+1}, \beta_j, \beta_{j+2}, \beta_{j+3}, ..., \beta_k]$;
$S := S \cup T$;
end;

Example

Consider the group G of degree 21 and order $3^4 \times 7^3$ which has a base

[1, 9, 8, 10, 2, 12]

and a strong generating set

s_1=(1,8,9)(2,11,15)(3,10,12)(4,14,19)(5,16,17)(6,21,20)(7,13,18),
s_2=(9,18,20)(12,19,17),
s_3=(10,21,11)(13,16,14),
s_4=(8,13,21)(10,14,16),
s_5=(2,6,3)(4,5,7), and
s_6=(12,20,15)(17,19,18).

The basic indices $|\Delta^{(i)}|$ are

27, 7, 7, 3, 3, 3.

Suppose we wish to interchange 1 and 9 in the base. Then we require generators for $\overline{G}^{(2)} = G_9$. Since $|9^G| = 21$, we calculate that $|\overline{\Delta}^{(2)}| = 7$. Initially,

$$\Delta = \{1\},\ \Gamma = \{2..8,\ 10..21\}, \text{ and } T = \{s_3, s_4, s_5, s_6\}.$$

Choosing $\gamma = 2$ from Γ gives

$$g_1 = s_1{}^2 \times s_2{}^2 \times s_6 \times s_1 = (1,2,3,4,6,5,7).$$

Then $9^{g_1^{-1}} = 9$, so g_2 is the identity, and we add

$$s_7 = g_1 = (1,2,3,4,6,5,7)$$

to T. Thus,

$$\Delta = \{1..7\},\ \Gamma = \{8,\ 10..21\}, \text{ and } T = \{s_3, s_4, s_5, s_6, s_7\}.$$

Since Δ has the correct size, we are finished. The new base is

[9, 1, 8, 10, 2, 12]

and the new strong generating set is

s_1=(1,8,9)(2,11,15)(3,10,12)(4,14,19)(5,16,17)(6,21,20)(7,13,18),
s_2=(9,18,20)(12,19,17),
s_3=(10,21,11)(13,16,14),
s_4=(8,13,21)(10,14,16),
s_5=(2,6,3)(4,5,7),
s_6=(12,20,15)(17,19,18), and
s_7=(1,2,3,4,6,5,7).

Note that s_2 is redundant in the new strong generating set. It was there to help generate G_1, which is no longer in the stabiliser chain.

Example

Consider now the interchange of 1 and 8, so that the new base will be

$$[9, 8, 1, 10, 2, 12].$$

We require the generators for $\overline{G}^{(3)} = G_{9,8}$. The orbit of 8 under $G^{(2)} = G_9$ is $\{8,10,11,13,14,16,21\}$, and so we calculate that $|\overline{\Delta}^{(3)}| = 7$. Initially,

$$\Delta = \{1\},\ \Gamma = \{2..7\}, \text{ and } T = \{s_3, s_5, s_6\}.$$

Choosing $\gamma = 2$ from Γ gives

$$g_1 = s_7 = (1,2,3,4,6,5,7).$$

Then $8^{g_1^{-1}} = 8$, so g_2 is the identity, and we add s_7 to T. Thus,

$$\Delta = \{1..7\},\ \Gamma = \text{empty, and } T = \{s_3, s_5, s_6, s_7\}.$$

Since Δ has the correct size, we are finished.

The new strong generating set is the same as the old. By checking not only $S^{(j+2)}$ but also $S^{(j)}$ for elements that fix β_{j+1}, we could have saved ourselves the trouble of duplicating permutations already in S. This simple improvement is used in implementations. We leave it as an exercise for the reader to make the necessary modifications to Algorithm 3.

Analysis of Interchanging Base Points

To analyse Algorithm 3, let

$$\begin{aligned} N_\gamma &= \text{number of choices of } \gamma \in \Gamma \text{ made, and} \\ N_{gen} &= \text{number of generators } g_2 \times g_1 \text{ added to } T. \end{aligned}$$

Then

$$\begin{aligned} N_\gamma &\le |\Delta^{(j)}| - 1, \text{ and} \\ N_{gen} &\le |\overline{\Delta}^{(j+1)}| - 1. \end{aligned}$$

Also note that the size of $\gamma^{<T>}$ in the **else**-clause is bounded by

$$|\Delta^{(j)}| - |\overline{\Delta}^{(j+1)}|.$$

The cost of initialisation is bounded by

$$\begin{array}{ll} |\overline{\Delta}^{(j)}| \times \left[2 \times |S^{(j)}| + 3 \right] + 3 \times |\Omega| & \text{for computing } \beta_{j+1}{}^{G^{(j)}} \\ + \ |\Omega| + 2 & \text{for the assignment } \Gamma := \Delta - \{\beta_j, \beta_{j+1}\} \\ + \ |\Omega| & \text{for the assigment } \Delta := \{\beta_j\}. \end{array}$$

The **while**-loop is executed N_γ times. The number of those iterations that execute the **then**-clause is N_{gen}, while $N_\gamma - N_{gen}$ iterations execute the **else**-clause. Hence, the total cost of the **while**-loop is bounded by

N_γ times

$$2 \times |\Delta^{(j)}| \times \left[\begin{array}{l} |\Omega| + 1 \\ + 2 \end{array} \right] \quad \begin{array}{l} \text{to form } g_1 \\ \text{to test the condition } \beta_{j+1}{}^{g_1^{-1}} \in \Delta^{(j+1)} \end{array}$$

plus N_{gen} times

$$2 \times |\Delta^{(j+1)}| \times \left[\begin{array}{l} |\Omega| + 1 \\ + 2 \times |\Omega| \end{array} \right] \quad \begin{array}{l} \text{to form } g_2 \\ \text{to form the product} \end{array}$$

$$+ \ |\overline{\Delta}^{(j+1)}| \times \left[\begin{array}{l} 2 \times |T| + 1 \\ + |\Omega| \end{array} \right] + |\Omega| \quad \begin{array}{l} \text{to compute } \Delta \\ \text{to form set difference} \end{array}$$

plus $N_\gamma - N_{gen}$ times

$$\left[|\Delta^{(j)}| - |\overline{\Delta}^{(j+1)}| \right] \times \left[\begin{array}{l} 2 \times |T| + 1 \\ + |\Omega| \end{array} \right] + |\Omega| \quad \begin{array}{l} \text{to form } \gamma^{<T>} \\ \text{to form set difference.} \end{array}$$

The final addition to the cost is to compute the Schreier vector of $\overline{\Delta}^{(j+1)}$. The calculation of

$$\beta_{j+1}{}^{G^{(j)}}$$

already gives a Schreier vector of $\overline{\Delta}^{(j)}$. This adds

$$|\overline{\Delta}^{(j+1)}| \times \left[2 \times |T| + 3 \right] + 3 \times |\Omega|$$

to the total.

The actual total is not enlightening. Bounding the size of the generating sets by $|\Omega|^2$ gives the order of Algorithm 3 as

$$O(\,|\Omega|^4\,).$$

In practice, N_{gen} is seldom larger than 1 or 2 and N_γ is generally quite small.

Removing Redundant Strong Generators

We saw in the first example of interchanging base points, that the strong generating set may contain redundancies. In this section we present a quick method of detecting and removing some redundancies. That is, we will look at each strong generator (in some order) and see if it is in the group generated by the previous strong generators. This check is easy if we work up from the bottom of the stabiliser chain. Working from the bottom also guarantees that the result is a strong generating set.

Suppose $T \subseteq S^{(i)}$ and $G^{(i+1)} \leq <T>$. Then we are concerned that T generates $G^{(i)}$. Suppose that $g \in G^{(i)}$. Then $g \in <T>$ if and only if $\beta_i{}^g \in \beta_i{}^{<T>}$, because $<T>$ contains $G^{(i+1)}$. This simplifies the redundancy test. The algorithm is Algorithm 4.

Algorithm 4 : Removing Some Redundancies

Input : a base $B = [\beta_1, \beta_2, ..., \beta_k]$;
a strong generating set S of a group G;

Output : a subset T of S that is also a strong generating set of G relative to B;

```
begin

  T := empty;

  for i := k downto 1 do
    for each generator s in S^(i) - S^(i+1) do
      if β_i^s ∉ β_i^<T> then
        T := T ∪ {s};
      end if;
    end for;
  end for;

end;
```

The cost of the algorithm is essentially $|T|$ orbit calculations.

Empirical evidence shows that the interchanges of base points introduces many redundancies. The above algorithm applied to the strong generating set produced by "random" base changes (involving several interchanges) often reduces the number of strong generators by 30-60%. (For these figures, the generators of $S^{(i)} - S^{(i+1)}$ were traversed in the reverse order to which they were originally added to the strong generating set.)

Conjugation and the Complete Algorithm

There are two remarks that lead to the work in this section. The first is that we have not used new bases of type (4) - $\bar{B} = [\beta_1^g, \beta_2^g, ..., \beta_k^g]$, where $g \in G$. The second is that the number of interchanges required to position the first point(s) in the new base is(are) generally larger than the number required to position the remaining base points. The first remark can help alleviate the second problem, since the groups we treat are often transitive.

Let B be the old base and $\bar{B}$ be the new base. From chapter 10, we know how to decide whether there exists an element mapping some initial segment of the old base to an initial segment of the new base. We can determine such an element as well. By taking such an element g, and applying a type (4) transformation we have positioned the initial segment of the new base. Interchanges will complete the task.

Algorithm 5 presents the complete base change. Empirical evidence indicates that, on average, the algorithm using conjugation is three times faster than an algorithm using only interchanges.

Algorithm 5 : Complete Base Change

Input : a base $B = [\beta_1, \beta_2, ..., \beta_k]$ and a strong generating set S of a group G;
a sequence $B' = [\beta'_1, \beta'_2, ..., \beta'_{k'}]$ to form an initial segment of the new base;

Output : a base $\overline{B}$ for G that is an extension (or initial segment) of B' and
a strong generating set $\overline{S}$ relative to $\overline{B}$;

begin

(*find a conjugating element g*)
$i := 0$; $g := id$; $more := \beta'_1 \in \Delta^{(1)}$;
while *more* **do**
  $i := i + 1$; $g := trace(\beta'^{g^{-1}}_i, v^{(i)}) \times g$;
  $more := (i+1) \leq \min(k, k')$;
  if *more* **then** $more := \beta'^{g^{-1}}_{i+1} \in \Delta^{(i+1)}$; **end if**;
end while;

(*conjugate by a nontrivial element*)
if $g \neq id$ **then**
  $B := B^g$; $S := S^g$;
  translate the basic orbits and Schreier vectors by g;
end if;

(*transpose remaining points into position*)
for $i := i+1$ **to** k' **do**

  if $\beta'_i \in B$ **then**
    pos := position of β'_i in B;
  else
    append β'_i to B;   *pos* := length of B;
  end if;

  while $pos \neq i$ **do**
    interchange($B, S, pos-1$); $pos := pos - 1$;
  end while;

end for;

delete redundant base points from the end of B;

(*return results*)
$\overline{B} := B$; $\overline{S} := S$;

end;

Example

Consider the group G of degree 21 and order $3^4 \times 7^3$. Let the original base be

[1, 9, 8, 10, 2, 12].

A base beginning with 9,8 can be obtained by conjugating by $g = s_1{}^2$. The conjugated base is

[9, 8, 1, 3, 15, 10]

and the corresponding strong generating set is

s_1=(1,8,9)(2,11,15)(3,10,12)(4,14,19)(5,16,17)(6,21,20)(7,13,18),
s_2=(8,13,21)(10,14,16),
s_3=(2,3,6)(4,7,5),
s_4=(1,7,6)(4,5,11),
s_5=(12,15,20)(17,18,19), and
s_6=(10,21,11)(13,16,14).

Summary

The base change algorithm is an efficient means of determining a strong generating set relative to a base, provided that some base and strong generating set for the group is known.

Exercises

(1/Easy) Modify Algorithm 3 to initially include in T not only $S^{(j+2)}$ but also those generators of $S^{(j)}$ which fix β_{j+1}.

(2/Easy) Change the base of the group G of degree 21 used in the examples from

[1, 9, 8, 10, 2, 12]

to

[9, 8, 1, 10, 2, 12]

and then to

[1, 2, 8, 9, 10, 12.]

Delete redundant strong generators.

(3/Moderate) For Algorithm 4 develop an algorithm that translates an orbit or Schreier vector by an element g. That is, given an element g of the group, an orbit $\Delta^{(i)}$ or a Schreier vector $v^{(i)}$ of $G^{(i)}$ relative to the generators $S^{(i)}$ compute an orbit $\overline{\Delta}^{(i)}$ or Schreier vector $\overline{v}^{(i)}$ of $\overline{G}^{(i)} = \left[G^{(i)} \right]^g$ relative to the generators $\left[S^{(i)} \right]^g$.

Do this in one scan of the orbit or Schreier vector.

Bibliographical Remarks

The base change algorithm is due to C. C. Sims, "*Determining the conjugacy classes of a permutation group*", **Computers in Algebra and Number Theory** (Proceedings of the Symposium on Applied Mathematics, New York, 1970), G. Birkhoff and M. Hall, Jr (editors), SIAM-AMS Proceedings, volume 4, American Mathematics Society, Providence, Rhode Island, 1971, 191-195. A much fuller description is given in C. C. Sims, "*Computation with permutation groups*", (Proceedings of the Second Symposium on Symbolic and Algebraic Manipulation, Los Angeles, 1971), S. R. Petrick (editor), Association of Computing Machinery, New York, 1971, 23-28. The use of conjugation is investigated in G. Butler, **Computational Approaches to Certain Problems in the Theory of Finite Groups**, Ph. D. Thesis, University of Sydney, 1980, leading to the complete base change algorithm presented in this chapter. This is also the source of our empirical evidence.

The algorithm for removing redundant strong generators was first introduced by Sims in the Los Angeles paper, and later improved by the author in his thesis.

Recently C.A. Brown, L. Finkelstein, and P.W. Purdom, Jr, "*A new base change algorithm for permutation groups*", SIAM Journal of Computing **18**, 5 (1989) 1037-1047, have shown that transposition of base points can be generalized to cyclic right shifts of base points. They present an algorithm and analyse it to be $O(|\Omega|^3)$.

Chapter 13. Schreier-Sims Method

Finally, we explain how to construct a base and strong generating set, given a set of generators. The method is founded on a result by Schreier, and was first developed by Sims. After presenting the original Schreier-Sims method, this chapter will discuss some variations.

Verifying Strong Generation

The "classical" viewpoint for the Schreier-Sims method is that we have a (partial) base B and a set S of elements of a group G, where S contains a generating set of G. We wish to verify that B is indeed a base, and that S is a strong generating set relative to B. If we discover that this is not so, then we wish to extend either B or S, or both, until it is true.

The simplest case is where S is the original set of generators and we choose some points $[\beta_1, \beta_2, \ldots, \beta_k]$ to form B, so that no generator fixes all k points.

Let

$$\begin{aligned} S^{(i)} &= \{ s \in S \mid s \text{ fixes } \beta_1, \beta_2, \ldots, \beta_{i-1} \}, \\ H^{(i)} &= \langle S^{(i)} \rangle, \text{ and} \\ G^{(i)} &= G_{\beta_1, \beta_2, \ldots, \beta_{i-1}}, \quad 1 \le i \le k+1. \end{aligned}$$

Hence, $H^{(k+1)} = \{id\}$. To verify that B is a base and S is a strong generating set, we need to show that

$$H^{(i)} = G^{(i)}, \text{ for all } i, \ 1 \le i \le k+1.$$

Once again, we will use an inductive approach, working from the bottom of the base to the top. In this way, we have a nice inductive hypothesis :

Hypothesis

Assume that B is a base for $H^{(i+1)}$ and that $S^{(i+1)}$ is a strong generating set of $H^{(i+1)}$ relative to B.

To prove the inductive hypothesis for i from the hypothesis for $i+1$, we need to show that

$$H^{(i)}{}_{\beta_i} = H^{(i+1)}.$$

Furthermore, we know that $H^{(1)} = G^{(1)} = G$. So, if we have proved that B is a base of $H^{(1)}$ and that $S = S^{(1)}$ is a strong generating set of $H^{(1)}$ relative to B, then we have the same result for G.

Not only does the inductive approach provide a neat proof of correctness, but it also allows us to assume we have a base and strong generating set of $H^{(i+1)}$. This allows us to easily answer questions involving $H^{(i+1)}$ and elements of G, such as membership.

An outline of an algorithm based on the above inductive hypothesis is presented as Algorithm 1.

Algorithm 1 : Outline of Schreier-Sims method

```
Input : a set S of generators of a group G;

Output : a base B for G;
         a strong generating set S of G relative to B;

procedure Schreier–Sims( var B : partial base; var S : set of elements; i : integer );
(* Assuming that B and S^(i+1) are a base and strong generating set
  for H^(i+1), produce a base and strong generating set for H^(i). *)
begin

  while H^(i)_βi ≠ H^(i+1) do

    find g ∈ H^(i)_βi − H^(i+1); find largest j such that g fixes β1, β2, ..., βj−1;

    add g to S;  (*actually to S^(i+1), S^(i+2), ..., S^(j) *)

    (*extend base, if necessary, so that no strong generator fixes all the base points*)
    if j = k+1 then
      find βj not fixed by g;
      add βj to B;
    end if;

    (*ensure we still have a base and strong generating set for H^(i+1) *)
    for level := j downto i+1 do
      Schreier–Sims( B, S, level );
    end for;
  end while;
end;
begin
  find points β1, β2, ..., βk so that no element of S fixes all of them;
  B := [β1, β2, ..., βk];
  for i := k downto 1 do
    Schreier–Sims( B, S, i );
  end for;
end;
```

The element $g \in H^{(i)}{}_{\beta_i} - H^{(i+1)}$ found in procedure *Schreier–Sims* will alter $S^{(j)}, S^{(j-1)}, \ldots, S^{(i+1)}$, and hence our assumptions about $H^{(j)}, H^{(j-1)}, \ldots, H^{(i+1)}$. Furthermore, if g fixes all the points presently in B then B must be extended.

Schreier Generators

The two open questions for the completion of the outline of the Schreier-Sims method are

1. How do we test $H^{(i)}{}_{\beta_i} = H^{(i+1)}$, and
2. How do we find an element $g \in H^{(i)}{}_{\beta_i} - H^{(i+1)}$?

The answer lies in the following result, which we will not prove. It is similar to the Loop Basis Theorem of Chapter 5.

Schreier's Lemma

Let $v^{(i)}$ be a Schreier vector of β_i under $H^{(i)}$ relative to the set $S^{(i)}$ of generators of $H^{(i)}$. Then $H^{(i)}{}_{\beta_i}$ is generated by

$$\{ \ trace(\gamma, v^{(i)}) \times s \times trace(\gamma^s, v^{(i)})^{-1} \ \mid \ \gamma \in \beta_i^{H^{(i)}}, s \in S^{(i)} \ \}.$$

The members of the above generating set are called *Schreier generators*.

The answer to both questions is to run through the Schreier generators - all

$$|\beta_i^{H^{(i)}}| \times |S^{(i)}|$$

of them - and test if they are in $H^{(i+1)}$. The membership test is straightforward because we have a base and strong generating set of $H^{(i+1)}$. If all the Schreier generators are in $H^{(i+1)}$, then $H^{(i)}{}_{\beta_i} = H^{(i+1)}$. If not, then any Schreier generator that is not in $H^{(i+1)}$ provides an element $g \in H^{(i)}{}_{\beta_i} - H^{(i+1)}$.

The fleshed out algorithm is presented as Algorithm 2.

Algorithm 2 : Using Schreier Generators in Schreier-Sims method

Input : a set S of generators of a group G;

Output : a base B for G;
a strong generating set S of G relative to B;

```
procedure Schreier-Sims( var B : partial base; var S : set of elements; i : integer );
(* Assuming that B and S^(i+1) are a base and strong generating set
  for H^(i+1), produce a base and strong generating set for H^(i).
  Note that H^(i) is invariant during the execution, and that
  the initial S^(i) is a set of generators of H^(i). *)
begin
  gen_set := S^(i);
  for each γ ∈ Δ^(i) do
    for each generator s ∈ gen_set do

      g := trace( γ, v^(i) ) × s × trace( γ^s, v^(i) )^-1;

      if g ∉ H^(i+1) then
        find largest j such that g fixes β1, β2, ..., βj-1;
        add g to S;  (*actually to S^(i+1), S^(i+2), ...,S^(j) *)
        (*extend base, if necessary, so that no strong generator
        fixes all the base points*)
        if j = k+1 then
          find βj not fixed by g;
          add βj to B;
        end if;

        (*ensure we still have a base and strong generating set for H^(i+1) *)
        for level := j downto i+1 do
          Schreier-Sims( B, S, level );
        end for;
      end if;

    end for;
  end for;
end;

begin
  find points β1, β2, ..., βk so that no element of S fixes all of them;
  B := [β1, β2, ..., βk];
  for i := k downto 1 do
    Schreier-Sims( B, S, i );
  end for;
end;
```

Example

We will execute Algorithm 2 using the symmetries of the projective plane of order two. The group is generated by a=(1,2,4,5,7,3,6), and b=(2,4)(3,5). We initially take $S=\{a, b\}$. We choose the initial partial base to be B = [1,2].

The first call to the procedure *Schreier–Sims* from the main algorithm is

$$Schreier\text{–}Sims\,(\,[1,2],\ \{a,b\},\ 2\,).$$

The relevant Schreier vector for forming the Schreier generators is

	1	2	3	4	5	6	7
$v^{(2)}$	0	0	0	b	0	0	0

The Schreier generators considered are

$id \times b \times b^{-1} = id$, for $\gamma = 2$, and

$b \times b \times id^{-1} = id$, for $\gamma = 4$.
Both are in $H^{(3)} = \{id\}$.

The next call to *Schreier–Sims* from the main algorithm is

$$Schreier\text{–}Sims\,(\,[1,2],\ \{a,b\},\ 1\,).$$

The relevant Schreier vector for forming the Schreier generators is

	1	2	3	4	5	6	7
$v^{(1)}$	0	a	a	a	a	a	a

The Schreier generators considered are

$id \times a \times a^{-1} = id$, for $\gamma = 1$;

$id \times b \times id^{-1} = b \in H^{(2)}$, for $\gamma = 1$;

$a \times a \times a^{-2} = id$, for $\gamma = 2$;

$a \times b \times a^{-2}$ = (2,6,3,7)(4,5) = g_1, for γ = 2; This is added to S, however, the base is not extended.

There is now a call to *Schreier–Sims* with i = 2 from the body of the procedure with i = 1. The call is

$$Schreier\text{–}Sims\,(\,[1,2],\ \{a,b,g_1\},\ 2\,).$$

The relevant Schreier vector for forming the Schreier generators is

	1	2	3	4	5	6	7
$v^{(2)}$	0	0	g_1	b	g_1	g_1	g_1

The Schreier generators considered are

$id \times b \times b^{-1} = id$, for $\gamma = 2$;

$id \times g_1 \times g_1^{-1} = id$, for $\gamma = 2$;

$g_1^2 \times b \times (b \times g_1)^{-1} = (4,7)(5,6) = g_2$, for $\gamma = 3$; This is added to S, and the base is extended by $\beta_3 = 4$. The call

$$Schreier\text{-}Sims\,(\,[1,2,4],\ \{a,b,g_1,g_2\},\ 3\,).$$

verifies that we have a base and strong generating set for $H^{(3)} = \langle g_2 \rangle$.

Back at $i = 2$, the processing of Schreier generators continues as follows:

$g_1^2 \times g_1 \times g_1^{-3} = id$, for $\gamma = 3$;

$b \times b \times id^{-1} = id$, for $\gamma = 4$;

$b \times g_1 \times (b \times g_1)^{-1} = id$, for $\gamma = 4$;

$(b \times g_1) \times b \times g_1^{-2} = g_2$, for $\gamma = 5$;

$(b \times g_1) \times g_1 \times b^{-1} = (4,5)(6,7) = g_3$, for $\gamma = 5$; This is added to S, however, the base is not extended. The call

$$Schreier\text{-}Sims\,(\,[1,2,4],\ \{a,b,g_1,g_2,g_3\},\ 3\,).$$

verifies that we have a base and strong generating set for $H^{(3)} = \langle g_2, g_3 \rangle$.

Back at $i = 2$, the processing of Schreier generators continues as follows:

$g_1 \times b \times g_1^{-1} = g_2 \times g_3$, for $\gamma = 6$;

$g_1 \times g_1 \times g_1^{-2} = id$, for $\gamma = 6$;

$g_1^3 \times b \times g_1^{-3} = g_2$, for $\gamma = 7$;

$g_1^3 \times g_1 \times id^{-1} = id$, for $\gamma = 7$.

Thus producing a base and strong generating set for $H^{(2)}$.

Back at $i = 1$, the processing of Schreier generators continues as follows:

$a^5 \times a \times a^{-6} = id$, for $\gamma = 3$;

$a^5 \times b \times a^{-3} = g_3 \times b \times g_1$, for $\gamma = 3$;

$a^2 \times a \times a^{-3} = id$, for $\gamma = 4$;

$a^2 \times b \times a^{-1} = g_1$, for $\gamma = 4$;

$a^3 \times a \times a^{-4} = id$, for $\gamma = 5$;

$a^3 \times b \times a^{-5} = (g_3 \times b \times g_1)^{-1}$, for $\gamma = 5$;

$a^6 \times a \times id^{-1} = id$, for $\gamma = 6$;

$a^6 \times b \times a^{-6} = g_3$, for $\gamma = 6$;

$a^4 \times a \times a^{-5} = id$, for $\gamma = 7$;

$a^4 \times b \times a^{-5} = g_3 \times g_1$, for $\gamma = 7$.

This completes the construction of a base and strong generating set. The result is a base

$$[1,2,4]$$

and a strong generating set

a=(1,2,4,5,7,3,6), b=(2,4)(3,5),
g_1=(2,6,3,7)(4,5), g_2=(4,7)(5,6), and g_3=(4,5)(6,7).

Note that the element g_1 is redundant as a strong generator. Further note that the second call to the procedure *Schreier–Sims* with $i = 2$ rechecked the Schreier generators corresponding to $\gamma = 2$ and 4 and generator b. The second call to the procedure *Schreier–Sims* with $i = 3$ rechecked the Schreier generators corresponding to $\gamma = 4$ and 7 and generator g_2.

Avoid Rechecking Schreier Generators

During the example, Algorithm 2 calls procedure *Schreier–Sims* with $i = 2$ twice. Each time the complete set of Schreier generators is checked for membership in $H^{(3)}$. At the first call

$$H^{(2)} = \langle b \rangle, \quad H^{(3)} = \{id\}, \quad \Delta^{(2)} = \{2,4\},$$

and the Schreier vector $v^{(2)}$ is

	1	2	3	4	5	6	7
$v^{(2)}$	0	0	0	b	0	0	0

A subscript 1 will distinguish these values. Thus, we will speak of $H^{(2)}{}_1$, $H^{(3)}{}_1$, $\Delta^{(2)}{}_1$, and $v^{(2)}{}_1$.

We can arrange for the Schreier vector $v^{(2)}$ to be extended whenever $H^{(2)}$ is extended. So the second call to the procedure *Schreier–Sims* has

$$H^{(2)} = \langle b, g_1 \rangle, \quad H^{(3)} = \{id\}, \quad \Delta^{(2)} = \{2,3,4,5,6,7\},$$

and the Schreier vector $v^{(2)}$ is

	1	2	3	4	5	6	7
$v^{(2)}$	0	0	g_1	b	g_1	g_1	g_1

A subscript 2 will distinguish these values. Thus, we will speak of $H^{(2)}{}_2$, $H^{(3)}{}_2$, $\Delta^{(2)}{}_2$, and $v^{(2)}{}_2$.

The extension of $v^{(2)}{}_1$ to $v^{(2)}{}_2$ is important. It allows us to claim that

$$trace(\gamma, v^{(2)}{}_2) = trace(\gamma, v^{(2)}{}_1), \text{for all } \gamma \in \Delta^{(2)}{}_1$$

and that the Schreier generator

$$trace(\gamma, v^{(2)}{}_2) \times s \times trace(\gamma^s, v^{(2)}{}_2)^{-1}$$
$$= trace(\gamma, v^{(2)}{}_1) \times s \times trace(\gamma^s, v^{(2)}{}_1)^{-1}$$

for all $\gamma \in \Delta^{(2)}{}_1$ and all generators s of $H^{(2)}{}_1$. As the first call to *Schreier–Sims* (with i=2) has verified that these Schreier generators are in $H^{(3)}$, there is no need for the second call to recheck this fact. Even if $H^{(3)}$ changes value, this is so, because the only possible change to $H^{(3)}$ is for $H^{(3)}$ to be extended.

The argument generalizes to show that, provided the Schreier vectors are calculated by extending their previous value, a call to *Schreier–Sims* with the value i does not need to recheck the Schreier generators considered by the previous calls the *Schreier–Sims* with the same value of i.

Algorithm 3 avoids the rechecking of Schreier generators. A further parameter T is introduced for the procedure. The parameter T is the subset of the generators $S^{(i)}$ that lie outside the previous value of $H^{(i)}$. That is, the generators s whose Schreier generators

$$trace(\gamma, v^{(i)}) \times s \times trace(\gamma^s, v^{(i)})^{-1}$$

for γ in the previous value of $\Delta^{(i)}$ are not yet known to lie in $H^{(i+1)}$.

Algorithm 3 : Schreier-Sims Method not rechecking Schreier Generators

Input : a set S of generators of a group G;

Output : a base B for G;
a strong generating set S of G relative to B;

procedure *Schreier–Sims*(**var** B : partial base; **var** S : set of elements;
i : integer; T : set of elements);
(* Assuming that B and $S^{(i+1)}$ are a base and strong generating set for $H^{(i+1)}$, produce a base and strong generating set for $H^{(i)}$.

T is the set of additional generators in $S^{(i)}$ since the previous call to the procedure with the present value of i.

Assume that a base and strong generating set of $<S^{(i)}-T>$, (the previous value of $H^{(i)}$), are included in B and S.

The present value of $v^{(i)}$ must be an extension of the previous value. *)

```
begin
  current_gens := S^(i); old_gens := S^(i) - T;
  old_Δ := β_i^<old_gens>; (*previous value of Δ^(i) *)

  for each γ ∈ Δ^(i) do

    if γ ∈ old_Δ then
      gen_set := T;
    else
      gen_set := current_gens;
    end if;

    for each generator s ∈ gen_set do
      g := trace( γ, v^(i) ) × s × trace( γ^s, v^(i) )^-1;

      if g ∉ H^(i+1) then
        find largest j such that g fixes β_1, β_2, ..., β_{j-1};
        add g to S;  (*actually to S^(i+1), S^(i+2), ..., S^(j) *)
        (*extend base, if necessary, so that no strong generator
        fixes all the base points*)
        if j = k+1 then
          find β_j not fixed by g;
          add β_j to B;
        end if;
        (*ensure we still have a base and strong generating set for H^(i+1) *)
        for level := j downto i+1 do
          Schreier-Sims( B, S, level, {g} );
        end for;
      end if;
    end for;

  end for;
end;

begin
  find points β_1, β_2, ..., β_k so that no element of S fixes all of them;
  B := [β_1, β_2, ..., β_k];
  for i := k downto 1 do
    Schreier-Sims( B, S, i, S^(i) );
  end for;
end;
```

Stripping Schreier Generators

Let us take a closer look at testing $g \in H^{(i+1)}$. The test attempts to express g as

$$g = u_k \times u_{k-1} \times \cdots \times u_{i+1}$$

for suitable $u_j \in H^{(j)}$ determined from the Schreier vectors. If the test fails, it is because some suitable u_l, $k \leq l \leq i+1$, cannot be found. Thus

$$g = \bar{g} \times u_{l-1} \times u_{l-2} \times \cdots \times u_{i+1}$$

where $u_j \in H^{(j)}$ and $\bar{g} \notin H^{(l)}$. We call $\bar{g}$ the *residue* of testing $g \in H^{(i+1)}$. If $g \in H^{(i+1)}$ then the residue is the identity. The process of determining the residue is called *stripping*.

When g is added to S, and procedure *Schreier–Sims* is called at level $i+1$, it must eventually extend $H^{(l)}$ by some generator related to $\bar{g}$. However, $\bar{g}$ and g are not independent. In fact, g will be a redundant generator of $H^{(i+1)}$ once $\bar{g}$ is added to S. So, why not just add $\bar{g}$ to S in the first instance, and forget about adding g. This not only leads to smaller strong generating sets, but also extends $H^{(l)}$ much sooner. This idea is used in Algorithm 4.

Algorithm 4 : Schreier-Sims Method stripping Schreier Generators

Input : a set S of generators of a group G;

Output : a base B for G;
a strong generating set S of G relative to B;

procedure *Schreier–Sims*(**var** B : partial base; **var** S : set of elements;
i : integer; T : set of elements);
(* Assuming that B and $S^{(i+1)}$ are a base and strong generating set for $H^{(i+1)}$, produce a base and strong generating set for $H^{(i)}$.

T is the set of additional generators in $S^{(i)}$ since the previous call to the procedure with the present value of i.

Assume that a base and strong generating set of $<S^{(i)} - T>$, (the previous value of $H^{(i)}$), are included in B and S.

The present value of $v^{(i)}$ must be an extension of the previous value. *)

```
begin
  current_gens := S^(i); old_gens := S^(i) - T;
  old_Δ := β_i^<old_gens>; (*previous value of Δ^(i) *)

  for each γ ∈ Δ^(i) do

    if γ ∈ old_Δ then
      gen_set := T;
    else
      gen_set := current_gens;
    end if;

    for each generator s ∈ gen_set do

      g := trace( γ, ν^(i) ) × s × trace( γ^s, ν^(i) )^-1;

      if g ∉ H^(i+1) then

        ḡ := residue of testing g ∈ H^(i+1);
        j := level l where testing stopped;  (*may be k+1*)

        add ḡ to S;  (*actually to S^(i+1), S^(i+2), ...,S^(j) *)
        (*extend base, if necessary, so that no strong generator
        fixes all the base points*)
       if j = k+1 then
          find β_j not fixed by ḡ;
          add β_j to B;
        end if;
        (*ensure we still have a base and strong generating set for H^(i+1) *)
        for level := j downto i+1 do
          Schreier-Sims( B, S, level, {ḡ} );
        end for;
      end if;
    end for;

  end for;
end;

begin
  find points β1, β2, ..., βk so that no element of S fixes all of them;
  B := [β1, β2, ..., βk];
  for i := k downto 1 do
    Schreier-Sims( B, S, i, S^(i) );
  end for;
end;
```

Variations of the Schreier-Sims Method

This section will discuss some variations of the Schreier-Sims method. They are all variations on Algorithm 4. They vary sometimes in the amount of information known at the start - for example, a base may be known - but mostly they differ in strategies to save space and time.

Original Schreier-Sims Method: The original algorithm that Sims devised used coset representatives rather than Schreier vectors. While being more space-consuming, empirical evidence indicates it is a factor of three faster.

Random Schreier-Sims Method: If $H^{(i)}{}_{\beta_i} \neq H^{(i+1)}$ then $H^{(i+1)}$ is a proper subgroup of $H^{(i)}{}_{\beta_i}$. Therefore, it has index at least two. This means that the probability of finding an element $g \in H^{(i)}{}_{\beta_i} - H^{(i+1)}$ is *at least* one half.

The random Schreier-Sims method tests $H^{(i)}{}_{\beta_i} = H^{(i+1)}$ by considering a number of (hopefully) random elements g of G and testing whether $g \in H^{(1)}$. If the residue $\overline{g}$ is not trivial, then $\overline{g}$ is a new strong generator. If t consecutive random elements of G are stripped to the identity then the probability that B and S are a base and strong generating set is $1-2^{-t}$.

Schreier-Todd-Coxeter-Sims Method: This method not only constructs a base and strong generating set, but also constructs a set of defining relations for the group G involving all the strong generators. The Todd-Coxeter algorithm can compute the index of $H^{(i+1)}$ in $H^{(i)}$, provided sufficient relations are known. The index should be $|\Delta^{(i)}|$. If there are insufficient relations, or the index is too large, the output of the Todd-Coxeter algorithm indicates which words in the generators $S^{(i)}$ it believes are the coset representatives of $H^{(i+1)}$ in $H^{(i)}$. Checking the image of β_i under these words will discover two words w_1 and w_2 that actually represent the same coset. Let $g = w_1 \times w_2^{-1}$. Then $g \in H^{(i)}{}_{\beta_i}$. Either $g \in H^{(i+1)}$ and we obtain another relation, or $g \notin H^{(i+1)}$ and we obtain a new strong generator. This process iterates until the Todd-Coxeter algorithm does compute the index $|\Delta^{(i)}|$.

Extending Schreier-Sims Method: Given a base B and strong generating set S of a group G and an element $g \notin G$, we find a base and strong generating set of $<G, g>$. This is simply a call

$$Schreier\text{–}Sims(B, S \cup \{g\}, 1, \{g\})$$

to the procedure of Algorithm 4.

This task is frequently used in other algorithms, for example, those algorithms of chapters 4 and 6. In most contexts we are extending a subgroup of a group for which we know a base. This not only gives us a base for the extended subgroup, but also allows the formation of Schreier generators and their stripping to be done in terms of base images. A complete permutation is required only in the few cases which lead to a new strong generator. This variation is called the **known base Schreier-Sims method.**

Summary

The Schreier-Sims method produces a base and strong generating set of a group given by generators. It does this by verifying that all the Schreier generators can be expressed in terms of coset representatives.

There are several variations on the Schreier-Sims method.

Exercises

(1/Moderate) The Schreier-Sims methods are very tedious to perform by hand for all but the smallest examples. Execute Algorithm 3 on the symmetries of the square, the symmetric group of degree 4, and the automorphism group of Petersen's graph.

(2/Moderate) Modify Algorithm 3 to use the sets $U^{(i)}$ of coset representatives rather than the Schreier vectors. Note that the sets $U^{(i)}$ must be *extended* when $H^{(i)}$ is extended, for the same reason that the Schreier vectors had to be extended.

(3/Moderate) For the random Schreier-Sims method, how would you determine a "random" element?

Bibliographical Remarks

The idea for the Schreier-Sims method is first presented in C. C. Sims, "*Computational methods in the study of permutation groups*", **Computational Problems in Abstract Algebra**, (Proceedings of a conference, Oxford, 1967), John Leech (editor), Pergamon, Oxford, 1970, 169-183. The paper indicates that Sims had implemented the method. The method is more fully presented in C. C. Sims, "*Computation with permutation groups*", (Proceedings of the Second Symposium on Symbolic and Algebraic Manipulation, Los Angeles, 1971), S. R. Petrick (editor), Association of Computing Machinery, New York, 1971, 23-28.

An early implementation is described in an unpublished manuscript : Karin Ferber, "*Ein Program zur Bestimmung der Ordnung grosser Permutationsgruppen*", Kiel, 1967, 8 pages. Ferber's implementation regarded the basic transversals $U^{(i)}$ as the generators $S^{(i)} - S^{(i+1)}$, and worked top-down rather than bottom-up. The transversals were used to limit the size of the strong generating set one usually gets when working top-down.

Another early implementation is described in J. S. Richardson, **GROUP : A Computer System for Group-Theoretic Calculations**, M. Sc. Thesis, University of Sydney, 1973. This thesis also suggests the extending Schreier-Sims method.

The Schreier-Todd-Coxeter-Sims method is due to Sims in an unpublished manuscript : C. C. Sims, "*Some algorithms based on coset enumeration*", Rutgers University, 1974. Sims had an experimental APL implementation of the method. In 1975, J. S. Leon produced a fullscale implementation of the algorithm and extensively investigated its performance. His work is described in J. S. Leon, "*On an algorithm for finding a base and strong generating set for a group given by generating permutations*", Mathematics of Computation **35**, 151 (1980) 941-974. Leon also develops the random Schreier-Sims method in this paper.

The author implemented the extending Schreier-Sims method in 1975 and the Schreier-Todd-Coxeter-Sims method in 1978. This work is described in G. Butler, **Computational Approaches to Certain Problems in the Theory of Finite Groups**, Ph. D. Thesis,

University of Sydney, 1980, along with uses of the extending Schreier-Sims method. The extending Schreier-Sims method and some of its uses are also described in G. Butler and J. J. Cannon, "*Computing in permutation and matrix groups I : Normal closure, commutator subgroup, series*", Mathematics of Computation **39**, 160 (1982) 663-670.

An analysis of (essentially Ferber's implementation of) the Schreier-Sims method was first presented by M. Furst, J. Hopcroft, and E. Luks, "*Polynomial-time algorithms for permutation groups*", (Proceedings of the IEEE 21st Annual Symposium on the Foundations of Computer Science, October 13-15, 1980), 36-41, who also analyse some other group-theoretic algorithms.

Some references for the Todd-Coxeter algorithm are J. A. Todd and H. S. M. Coxeter, "*A practical method for enumerating cosets of a finite abstract group*", Proceedings of the Edinburgh Mathematical Society (2) **5** (1937) 26-34; J. J. Cannon, L. A. Dimino, G. Havas, and J. M. Watson, "*Implementation and analysis of the Todd-Coxeter algorithm*", Mathematics of Computation, **27** (1973) 463-490; and J. Neubüser, "*An elementary introduction to coset table methods in computational group theory*", **Groups-St Andrews 1981**, C. M. Campbell and E. F. Robertson (editors), London Mathematical Society Lecture Notes Series **71**, Cambridge University Press, Cambridge, 1982, 1-45.

Chapter 14. Complexity of the Schreier-Sims Method

In this chapter we present some variations of the Schreier-Sims method together with analyses of their complexity. The final result is an $O(|\Omega|^5)$ bound.

Furst, Hopcroft, and Luks

The view taken by Furst, Hopcroft, and Luks is that the base of the group is $[1,2,..|\Omega|]$ and that the algorithm is constructing a table $T[1..|\Omega|, 1..|\Omega|]$ of permutations, where $T[i,j]$ is an element of the group G which fixes 1,2,..,i-1 and maps i to j. If no such element exists, then the entry is empty. Hence, each element along the diagonal can be taken to be the identity element. The non-empty elements of the i-th row form a set of coset representatives $U^{(i)}$ of $G^{(i+1)}$ in $G^{(i)}$.

During the algorithm the table data structure is just required to have $T[i,j]$ empty, or an element of G which fixes 1,2,...,i-1 and maps i to j. For such tables we can still define the set $U^{(i)}$ of non-empty entries of the i-th row. They define *closure* (T) to be the set of products $\{ u_r u_{r-1} \cdots u_1 \mid u_i \in U^{(i)} \}$, and define a table to be *complete* if its closure is G.

The *canonical representative* of an element $g \in G$ relative to a table T is the product $u_r u_{r-1} \cdots u_1$, where $u_i \in U^{(i)}$, such that $g = u_r u_{r-1} \cdots u_1$. Not all elements may have canonical representatives.

Lemma 1

A table T is complete if and only if both

(a) every generator of G has a canonical representative, and

(b) every product of (non-empty) entries of T has a canonical representative.

The procedure *sift* of Algorithm 1 will compute the canonical representative of an element if it exists. It is essentially the membership algorithm. If a canonical representative does not exist, then the table T is updated by replacing an empty entry by the residue of the stripping process.

Algorithm 1 : Sift

Input: a (not necessarily complete) table T for the group G;
    an element $g \in G$;
Output: a modified table T such that $g \in closure(T)$;

```
procedure sift( g );
begin
  i := 1;
  while i ≤ |Ω| and T[i,i^g] is not empty do
    g := g × T[i, i^g]^-1;
    i := i+1;
  end while;
  if T[i,i^g] is empty then
    T[i,i^g] := g;
  end if;
end;
```

Lemma 2

Each invocation of *sift* is at worst $O(|\Omega|^2)$.

The algorithm then uses the characterization of Lemma 1 and repeated application of *sift* to force the table to be complete. One can take the non-empty entries of T to be a strong generating set for G relative to the given base.

Algorithm 2 : FHL Version of Schreier-Sims

```
Input: a set of generators of G;
Output: a complete table T for G;
begin
  initialize T so that the diagonal entries are the identity element
    and all other entries are empty;
  for each generator g of G do
    sift( g );
  end for;
  for each pair h, g of non-empty entries in T do
    sift( h × g );
  end for;
end.
```

Theorem 1

Algorithm 2 is $O(|\Omega|^6)$, provided that the size of the initial generating set is $O(|\Omega|^4)$.

Proof: There are at most $|\Omega|^2$ non-empty entries of T, so there are $O(|\Omega|^4)$ calls to sift.
□

Consider the algorithm's execution for the automorphism group of the projective plane of order 2. The number of calls to *sift* is 443 (or 198 if one disregards pairs where one element is the identity). This is many more than the number of Schreier generators constructed by the Schreier-Sims method.

Knuth

Knuth uses a similar data structure to Furst, Hopcroft, and Luks but turns the algorithm on its head to get a nice recurrence relation. He takes the base to be $[\ |\Omega|, |\Omega|-1, ..., 2, 1\ ]$. Then $G^{(i)}$ is the stabiliser of $[|\Omega|, |\Omega|-1,...,i+1]$. The entries of the table are such that $T[i,j]$ is an element of $G^{(i)}$ mapping i to j, or the entry is empty. Hence, the table is lower triangular with the identity element along the diagonal.

Define $T^{[j]}$ to be the upper left $j \times j$ subtable of T. All the elements in $T^{[j]}$ fix the points $\{j+1, j+2,..., |\Omega|\}$ and may be regarded as permutations of degree j.

Define $U^{(i)}$ to be the set of non-empty entries of the i-th row of T.

Define the *closure* of $T^{[j]}$ to be the set $\Gamma^{[j]}$ of products $u_1 u_2 \cdots u_j$, where $u_i \in U^{(i)}$.

Let S be a set of generators of G. (It will be extended to a strong generating set.) The data structure is the triple (S, T, Γ). The data structure is said to be *up to date of order m* if and only if $T^{[m]}$ is closed under multiplication.

If the table $T^{[m]}$ is closed under multiplication then $\Gamma^{[j]} = \langle S^{(j)} \rangle$, for all $j \le m$. This implies that we can test membership in $\Gamma^{[j]}$.

Lemma 3

Membership testing in $\Gamma^{[j]}$ can be done in $O(j^2)$ steps.

The algorithm for the Schreier-Sims consists of a pair of mutually recursive routines A_j and B_j.

Algorithm 3 : Knuth Version of Schreier-Sims

procedure $A_j(g)$;

Input: (S, T, Γ) up to date of order j;
 $g \in \langle S^{(j)} \rangle - \Gamma^{[j]}$;
Output: (S, T, Γ) up to date of order j and $g \in S$;

begin
 $S^{(j)} := S^{(j)} \cup \{g\}$;
 for each $h \in U^{(j)}$ **and** $s \in S^{(j)}$ **such that we do not know that** $h \times s \in \Gamma^{[j]}$ **do**
  $B_j(h \times s)$;
 end for;
end;

procedure $B_j(g)$;

Input: (S, T, Γ) up to date of order $j-1$;
 $g \in \langle S^{(j)} \rangle$;
Output: (S, T, Γ) up to date of order $j-1$ and $g \in \Gamma^{[j]}$;

begin
 $l := j^g$;
 if $T[j,l]$ is empty **then**
  $T[j,l] := g$;
 else if $g \times T[j,l]^{-1} \notin \Gamma^{[j-1]}$ **then**
  $A_{j-1}(g \times T[j,l]^{-1})$;

 end if;
end;

Theorem 2

The algorithm is correct.

The proof shows that the set $\Gamma^{[j]}$ is closed under multiplication when A_j terminates. The proof uses induction on j and on the length of words in $S^{(j)}$.

The result is obviously true for $j = 1$, as $\Gamma^{[1]}$ is the identity subgroup.

Assume the result is true for $j-1$. Let $h_1, h_2 \in \Gamma^{[j]}$. We need to show that $h_1 \times h_2 \in \Gamma^{[j]}$. We can write h_1 as $h \times u_1$ for some $h \in \Gamma^{[j-1]}$, $u_1 \in U^{(j)}$; and we can write h_2 as $s_1 \times s_2 \times \cdots \times s_q$ for some $s_i \in S^{(j)}$. The call to B_j guarantees that $u_1 \times s_1 \in \Gamma^{[j]}$, so we can write $u_1 \times s_1$ as $g_1 \times u_2$ for some $g_1 \in \Gamma^{[j-1]}$, $u_2 \in U^{(j)}$. The call to B_j guarantees that $u_2 \times s_2 \in \Gamma^{[j]}$, so we can write $u_1 \times s_2$ as $g_2 \times u_3$ for some $g_2 \in \Gamma^{[j-1]}$, $u_3 \in U^{(j)}$. Continuing in this fashion we can write $h_1 \times h_2$ as a product $h \times g_1 \times g_2 \times \cdots \times g_q \times u_q$, where each $g_i \in \Gamma^{[j-1]}$, $h \in \Gamma^{[j-1]}$, and $u_q \in U^{(j)}$. So $h_1 \times h_2 \in \Gamma^{[j]}$. □

Lemma 4

In executing A_j, the number of calls to B_j is bounded by $|U^{(j)}| \times |S^{(j)}|$.

At this point, Knuth noticed that each call to A_j increases the size of the subgroup $\Gamma^{[j]}$, so the question was how many such calls could be made. This is equivalent to asking how long can a proper chain of subgroups be in the symmetric group of degree j. By Lagrange's Theorem, the indices in such a chain must be divisors of the order $j!$ of the symmetric group. Define $\Theta(n)$ to be the number of prime divisors of n, counting multiplicity. Then the number of calls to A_j is less than or equal to $\Theta(j!)$, which is known to be $O(j\log\log(j))$.

Lemma 5

The cost of a sequence of calls to A_j is $O(j^5 \log log(j))$.

Proof: Let $C(j)$ be the total cost of calls to A_j. Then

$$C(j) = |U^{(j)}| \times |S^{(j)}| \times \left[j + j^2 \right] + C(j-1)$$

where the first factor j is the cost of forming $h \times s$, and the second factor j^2 is the cost of testing membership using B_j. The last term $C(j-1)$ is the cost of calls to A_{j-1} from B_j. So

$$C(j) \leq O(j^4 \log\log(j)) + C(j-1).$$

As $C(1) = 1$, we get that

$$C(j) \leq \sum_{i=1}^{j} O(i^4 \log\log(i))$$

which is $O(j^5 \log\log(j))$.□

We must still insist that the number of generators supplied as input be of a reasonable size, that is $O(|\Omega|^3)$, as we must check whether each of them is in $\Gamma^{[|\Omega|]}$.

Theorem 3

The total cost of the algorithm is $O(|\Omega|^5 \log log(|\Omega|))$.

Tighter bounds on the length of proper chains of subgroups is the symmetric group were proven by Babai.

Theorem 4 (Babai)

The length of the longest proper chain of subgroups in the symmetric group of degree j is $O(j)$.

Corollary 1

Knuth's algorithm is $O(|\Omega|^5)$.

Jerrum

Jerrum assumes that the base is [1,2,..., $|\Omega|$] and introduced a new data structure for his view of the Schreier-Sims method. This data structure, a complete labelled branching, has also been used in later algorithms computing other group theoretic information. His algorithm is $O(|\Omega|^5)$ and uses $O(|\Omega|^2)$ storage, as opposed to the $O(|\Omega^3)$ storage requirements of Knuth's table.

A *branching* (Ω, E) is a acyclic directed graph (with no loops or multiple edges) with vertices Ω and edges E such that every vertex has at most edge directed into it.

A *labelled branching* is a triple (Ω, E, σ) such that

1. (Ω, E) is a branching;
2. $\sigma : E \to S_\Omega$ is a labelling of the edges by permutations;
3. for all edges ij, the label σ_{ij} stabilises $1,2,...,i-1$; and
4. for all edges ij, the label σ_{ij} maps i to j.

For a directed path $P = i_0 i_1 i_2 .. i_t$, define the permutation σ_P to be the product

$$\sigma_{i_0 i_1} \times \sigma_{i_1 i_2} \times \cdots \times \sigma_{i_{t-1} i_t}$$

Hence, σ_P is an element mapping i_0 to i_t.

A *complete labelled branching* of G is a labelled branching where the edge labels are elements of G, and such that the set

$$U^{(i)} = \{ \sigma_P \mid P \text{ is a directed path with start vertex } i \}$$

is a right transversal of $G^{(i+1)}$ in $G^{(i)}$, for each i, $1 \le i \le |\Omega|$.

One can think of a labelled branching as containing the Schreier vectors for all levels of the stabiliser chain in a single data structure. Let us consider some examples of complete labelled branchings: one for the alternating group A_4, and one for the symmetries of the projective plane.

Figure 1: Complete labelled branching of A_4

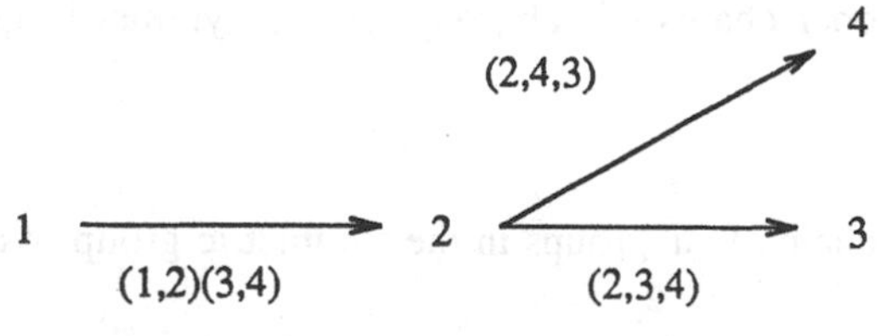

There are at most $|\Omega|$ edges in a labelled branching, so the storage requirement is $O(|\Omega|^2)$. Of course, it is important that such a data structure exists for each group, so we will show how one can construct it from an existing base and strong generating set and Schreier vectors. Then we will show how to construct it from scratch; that is, from a generating set. That is Jerrum's version of the Schreier-Sims method.

Figure 2: Complete labelled branching of symmetries of the projective plane

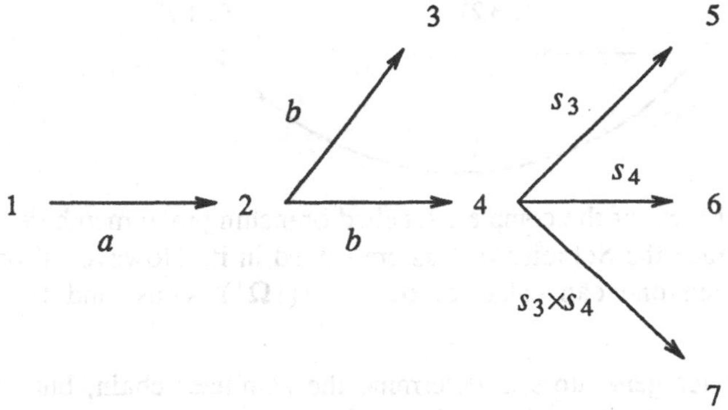

Construction from table: Suppose we have constructed the table of Furst, Hopcroft, and Luks for a group G. Then we find the last non-empty entry $T[i,j]$ in the j-th column. For this entry we define an edge ij with label $T[i,j]$. This defines a complete labelled branching of G.

Construction from transversals $U^{(i)}$: Suppose G has a base $[1,2,...,k]$ with strong generators and basic transversals $U^{(i)}$, for $1 \le i \le k$. Then the following algorithm constructs a complete labelled branching.

Algorithm 4 : Complete labelled branching

```
Input: a base [1,2,...k] for G;
   the basic transversals U^(i) of G;
Output: a complete labelled branching of G;

begin
  initialize the edge set E to be empty;
  for i := k to 1 do
    for each u in U^(i) do
      if the vertex i^u has no incoming directed edge then
        add an edge from i to i^u with label u to E;
      end if;
    end for;
  end for;
end.
```

Construction by enumeration: Suppose we have a base $[1,2,...,k]$ for G with a corresponding strong generating set and Schreier vectors. Then we can enumerate the elements in lexicographic order. For each point j, we find the first element $g \in G$ which maps some point $i<j$ to j. (This will guarantee that i is as small as possible.) We add the edge ij and take σ_{ij} to be g.

There is an equivalent view of complete labelled branchings which labels vertices rather than edges. Define the vertex label τ_j to be σ_P, where P is the longest path which ends at j. Then σ_{ij} is $\tau_j \times \tau_i^{-1}$, and if P is the path $i_0 i_1 i_2 \cdots i_t$, then σ_P is $\tau_{i_t} \times \tau_{i_0}^{-1}$. Hence, one can readily convert from the edge label view to the vertex label view.

Figure 3: Complete (vertex) labelled branching of A_4

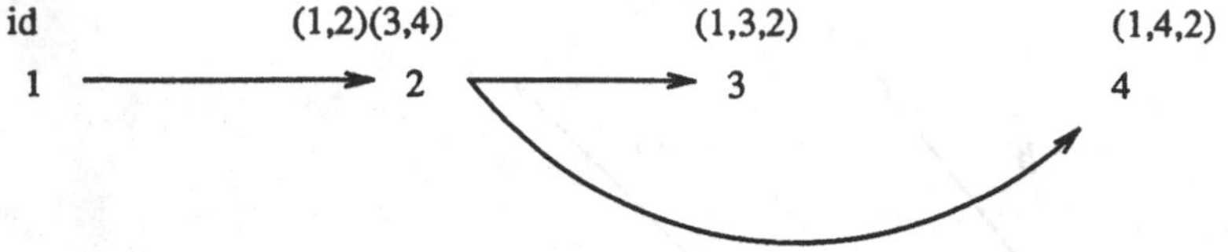

Note that if one uses the edge label view of the complete labelled branching then membership testing is $O(|\Omega|^3)$ as one must trace the Schreier vectors embedded in it. However, if one takes the vertex labelled view then one can calculate σ_P in $O(|\Omega|)$ steps, and hence membership testing is $O(|\Omega|^2)$.

Jerrum's algorithm uses the Schreier generators to determine the stabilizer chain, but can detect redundant Schreier generators at least good enough to reduce the number of generators from $O(|\Omega|^2)$ to less than $|\Omega|$. Hence, in his algorithm the number of generators for the stabilizer does not explode as he works down the chain.

Algorithm 5 : Jerrum Algorithm

Input: a set of generators for G;
Output: a complete labelled branching of G;
begin
  initialize the edge set E to be empty;
  for i := 1 **to** $|\Omega|$ **do**
    compute the orbit $\Delta^{(i)}$ and transversal $U^{(i)}$ from generators for $G^{(i)}$;
    augment E to include $U^{(i)}$;
    compute the $O(|\Omega|^2)$ Schreier generators of $G^{(i+1)}$;
    sift the Schreier generators to a set of generators of size less than $|\Omega|$;
  end for;
end.

procedure augment

Input: the edge set E and labelling;
  the level i;
  the basic orbit $\Delta^{(i)}$ and transversal $U^{(i)}$;
Output: an updated edge set E and labelling;

begin
  for each j in $\Delta^{(i)}$ - $\{i\}$ **do**
    if edge lj exists in E **then**
      remove the edge lj;
    end if;
    add edge ij to E with label from $U^{(i)}$ mapping i to j;
  end for;
end;

We still need to explain the sifting process. This is presented in Algorithms 6 and 7.

Algorithm 6 : Sift a set of generators

```
procedure sift( var T : set of generators );

Input: a set T of generators for a group G;
Output: a set T of generators of G of size less than |Ω|;

begin
  initialize Λ to be an empty labelled branching;
  for each g ∈ T do
    sift( g );
  end for;
  T := labels of Λ;
end;
```

Algorithm 7 : Sift an element

```
procedure sift( g : element );

Input: a labelled branching Λ (with value Λ0);
       an element g;
Output: an updated labelled branching Λ such that <labels(Λ)> = <labels(Λ0), g>;

begin
  j := first point moved by g;   l := j^g;
  while there exists a path from j to l in Λ and j < |Ω| do
    g := g×τ_l^-1×τ_j;
    j := first point moved by g;   l := j^g;
  end while;
  if j = |Ω| then
    return;
  end if;
  while there exists an edge ml in Λ with m > j do
    g := g×σ_ml^-1;   l := m;
  end while;
  if there exists an edge ml in Λ then
    add edge jl to Λ with label g;
    g := σ_ml×g^-1;
    remove edge ml from Λ;
    update the vertex labelling;
    sift( g );
  else
    add edge jl to Λ with label g;
    update the vertex labelling;
  end if;
end;
```

The aim of Algorithm 7 is to express the element g as σ_P for some path P. The first part of the algorithm is just a variation of the membership test. During the second **while**-loop, one can think of the branching as being modified as in Figure 4, although this modification is not actually done. It modifies the branching as in Figure 5 in the **if**-clause at the end, or by simply adding an edge in the **else**-clause.

Figure 4

Figure 5

The vital ingredient to prove the correctness of the sifting process is that the group $<$labels(B), $g>$ is invariant during *sift*(g). Once it is verified that each transformation of the labelled branching keeps the group invariant then it is clear the algorithm as a whole is correct.

Lemma 6

If the set T is $O(|\Omega|^2)$ then the cost of the call *sift*(T) is $O(|\Omega|^4)$.

If we ignore recursion, then the cost of each call *sift*(g) is $O(|\Omega|^2)$. However, each recursive call to *sift* occurs after

$$\text{edge length}(\Lambda) = \sum_{ij \in \Lambda} |j-i|$$

is reduced. The worst possible value of edge length is $|\Omega| \times (|\Omega|-1)/2$, so there are at most $O(|\Omega|^2)$ recursive calls to *sift* in total. □

Corollary 2

A complete labelled branching can be computed from a set of generators of size $O(|\Omega|)$ in time $O(|\Omega|^5)$.

Corollary 3

The group G has a strong generating set of size less than $|\Omega|$.

Let us consider the example of the symmetries of the projective plane. G is generated by a and b acting on $\{1,2,3,4,5,6,7\}$. For $i = 1$ we compute $U^{(1)} = \{ a^m \mid m = 0,1,...,6 \}$ and augment the empty labelled branching to obtain Figure 6 (omitting the labels).

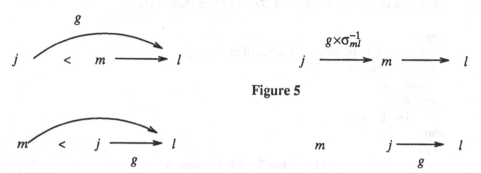

Figure 6

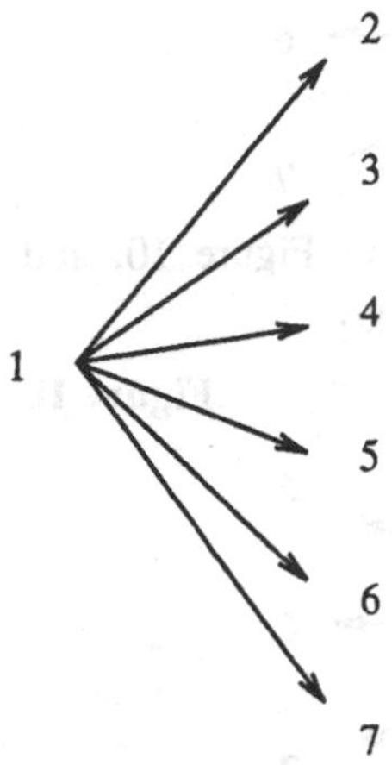

The non-trivial Schreier generators are b, (2,6,3,7)(4,5), (2,5,6)(3,4,7), (2,6,5)(3,7,4), (4,5)(6,7), (2,6)(3,7).

We call *sift* with this set and it processes each generator in turn. We start with an empty branching Λ. The sifting of the first three generators just adds an edge, so we have the branching in Figure 7.

Figure 7

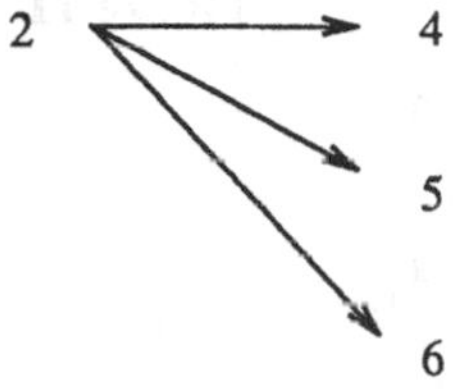

The fourth generator is tripped by σ_{26} to give g=(4,6)(5,7). The branching is modified to that in Figure 8, and we recursively sift $\sigma_{26} \times g^{-1}$ = (2,4,7)(3,5,6). This is stripped by σ_{24} to (4,7)(5,6) and a new edge is added to give the branching in Figure 9.

Figure 8

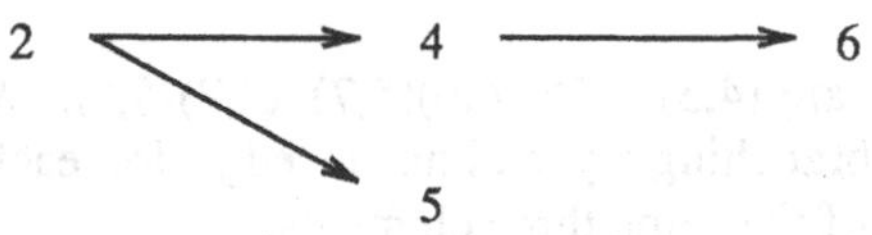

Figure 9

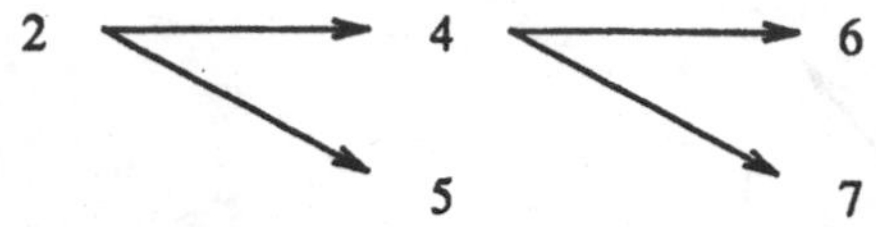

The fifth generator changes the branching to that in Figure 10, and we recursively sift $\sigma_{25} \times g^{-1} = (2,4,6)(3,5,7)$. This is stripped to the identity.

Figure 10

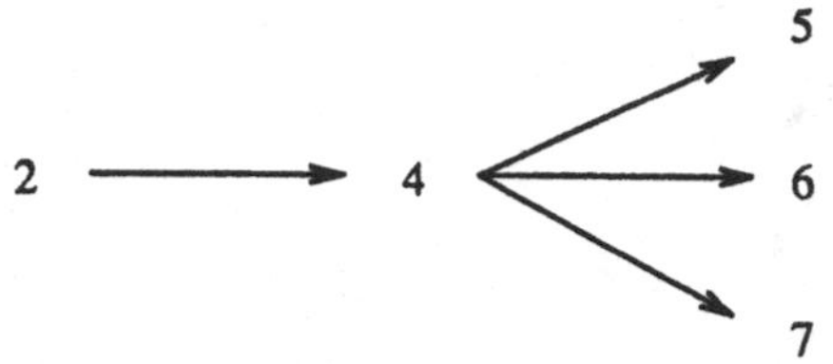

The sixth generator is stripped to the identity.

So we are finished. Our new generating set is
$\{ b, b\times(4,5)(6,7), b\times(4,6)(5,7), b\times(4,7)(5,6) \}$.

For $i = 2$, the Schreier vectors will give us $U^{(2)}$. The branching is augmented to that in Figure 11.

Figure 11

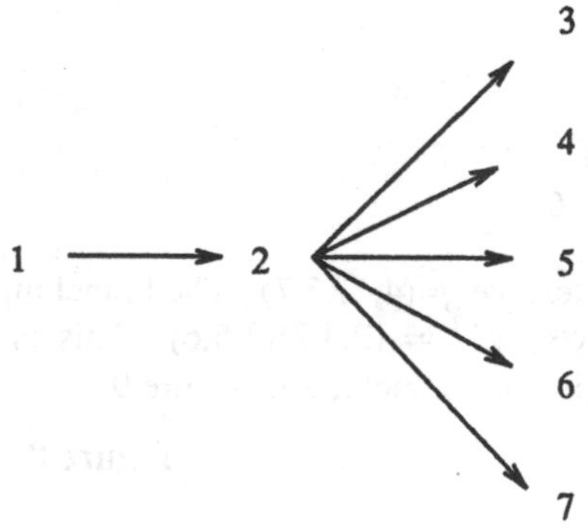

The non-trivial Schreier generators are (4,5)(6,7), (4,6)(5,7), (4,7)(5,6). We call *sift* with this set. It just builds up a labelled branching by adding an edge for each generator, so our resulting set of generators consists of the same three elements.

For $i = 3$, the set $U^{(3)}$ is constructed from the Schreier vectors. The branching is augmented to be that in Figure 2. The set of non-trivial Schreier generators is empty. Hence, we are finished. Our final strong generating set is
{ (4,5)(6,7), (4,6)(5,7), (4,7)(5,6), (2,3)(6,7), b, a }.
Note that the third and fourth are redundant.

Summary

We have presented some variations of the Schreier-Sims method which are simple enough to analyse. They all store actual permutations for the coset representatives. Jerrum's algorithm gives an $O(|\Omega|^5)$ time complexity and an $O(|\Omega|^2)$ space requirement.

Exercises

(1/Moderate) Execute Knuth's algorithm on the symmetries of the projective plane generated by a and b.

(2/Moderate) Which of the algorithms of this chapter and Algorithm 4 of the previous chapter would you expect to be fastest in practice? Why?

(3/Moderate) Repeat exercise 2 but modify Algorithm 4 of the previous chapter so it stores the coset representatives as permutations (and not in a Schreier vector).

(4/Difficult) Can the analyses of this chapter be carried over to Algorithm 4 of the previous chapter (the Schreier-Sims method)? Where does the difficulty lie?

Bibliographical Remarks

The first analysis of the Schreier-Sims method was presented by M. Furst, J. Hopcroft, and E. Luks, "*Polynomial-time algorithms for permutation groups*", (Proceedings of the IEEE 21st Annual Symposium on the Foundations of Computer Science, October 13-15, 1980), 36-41.

Knuth then presented his version of the Schreier-Sims method in an unpublished manuscript entitled "*Notes on efficient representation of perm groups*" in 1981. Jerrum's results were first presented in M. Jerrum, "*A compact representation for permutation groups*", (Proceedings of the IEEE 23rd Annual Symposium on the Foundations of Computer Science, 1982), 126-133, and later published in Journal of Algorithms **7** (1986) 60-78 in expanded form.

Since 1982 there has been progress in bounding the length of subgroup chains in the symmetric group by L. Babai, "*On the length of subgroup chains in the symmetric group*", Communications in Algebra **14** (1986) 1729-1736, who determined the $O(|\Omega|)$ bound on the length of chains, and minor progress in improving the complexity bounds of determining a base and strong generating set. L. Babai, E.M. Luks, and A. Seress, "*Fast management of permutation groups*", (Proceedings of the IEEE 29th Annual Symposium on the Foundations of Computer Science, October 24-26, 1988), 272-282, give an algorithm which is $O(|\Omega|^4 \log^c |\Omega|)$, where c is an undetermined constant. There has also been investigation of the parallel complexity of the problem by E.M. Luks and P. McKenzie, "*Fast parallel computation with permutation groups*", (Proceedings of the IEEE 26th Annual Symposium on the Foundations of Computer Science, 1985), 505-514, and L. Babai, E.M. Luks, and A. Seress, "*Permutation groups in NC*", (Proceedings of the 19th Annual ACM Symposium on Theory of Computing, May 25-27, 1987), 409-420.

Chapter 15. Homomorphisms

There are several classes of homomorphisms that arise naturally in the study of permutation groups. They are very useful in divide-and-conquer techniques for reducing a problem in a group G to smaller, less complicated groups. The aim of this chapter is to present methods of effectively computing with these naturally arising homomorphisms.

Homomorphisms

A *homomorphism* $\phi : G \rightarrow H$ between groups G and H is a map which commutes with the operations of multiplication and inversion, and which maps the identity element of G to the identity element of H. The group G is called the *domain* and the group H is called the *codomain*. For all elements $g_1, g_2 \in G$, we have

$$\phi(g_1 \times g_2) = \phi(g_1) \times \phi(g_2)$$
$$\phi(g_1^{-1}) = \phi(g_1)^{-1}$$

Hence, the homomorphism ϕ maps subgroups of G to subgroups of $\phi(G)$, and maps normal subgroups of G to normal subgroups of $\phi(G)$. The *image*, *im*ϕ, of the homomorphism ϕ is the subgroup $\phi(G) = \{ \phi(g) \mid g \in G \}$ of H. The *kernel*, *ker*ϕ, of the homomorphism ϕ is the subgroup $\{ g \in G \mid \phi(g) = identity \text{ of } H \}$ of G. Let T be a subset of elements of *im*ϕ. The *preimage*, $\phi^{-1}(T)$, of T is the set of all elements of G which map to an element of T. That is,

$$\phi^{-1}(T) = \{ g \in G \mid \phi(g) \in T \}$$

The preimage of a subgroup of *im*ϕ is a subgroup of G, and the preimage of a normal subgroup of *im*ϕ is a normal subgroup of G. The kernel of ϕ is the preimage of the identity element of H.

The kernel of ϕ is a normal subgroup N of G and the image of ϕ is isomorphic to the quotient group G/N. Every normal subgroup N of G give rise to a natural homomorphism

$$\phi : G \rightarrow G/N$$
$$g \mapsto g \times N$$

where an element is mapped to the coset of N which contains it.

A homomorphism may be defined in two ways. The first way is to give a formula or procedure which explicitly defines the image of each element of the group. For example, we can define the homomorphism ϕ between the symmetric group G of degree 4 and the cyclic group $H = < z >$ of order 2 by

$$\phi : G \mapsto H$$
$$g \mapsto z^{parity(g)}$$

where the parity of a permutation g is 0 if g can be expressed as a product of an even number of transpositions, and is 1 otherwise. The second way is to define the image of each generator s of G and allow the image of an element to be defined by expressing the element as a word in the generators. So if

$$g = s_1^{\varepsilon_1} \times \cdots \times s_t^{\varepsilon_t}$$

then the image is

$$\phi(g) = \phi(s_1)^{\varepsilon_1} \times \cdots \times \phi(s_t)^{\varepsilon_t}$$

So the homomorphism between the symmetric group of degree 4 and the cyclic group of order 2 can be defined by

$$\begin{aligned} \phi : G &\rightarrow H \\ (1,2,3,4) &\mapsto z \\ (1,2,3) &\mapsto identity \end{aligned}$$

Overview

There are several classes of homomorphisms that arise naturally in the study of permutation groups. Let G be a permutation group acting on a set Ω. Given a G-invariant subset Δ of Ω, the *transitive constituent homomorphism* sends each g in G to its restriction on Δ. If π is a system of imprimitivity (or block system) for G, the *blocks homomorphism* sends each g in G to the permutation it induces on π.

The aim is to present methods of effectively computing with these naturally arising homomorphisms. We shall take the notion of effective computation with a homomorphism $\phi: G \rightarrow \overline{G}$ to mean the ability to perform the following tasks.

1. Construct the image $im\phi$ of G.
2. Construct the kernel $ker\phi$.
3. Given an element g of G, construct $\phi(g)$.
4. Given an element h of $im\phi$, construct an element g of $\phi^{-1}(h)$.
5. Given a subgroup H of G, construct $\phi(H)$.
6. Given a subgroup $\overline{H}$ of $im\phi$, construct $\phi^{-1}(\overline{H})$.

Of course, to construct a subgroup of a permutation group will mean forming a base and strong generating set of the subgroup. Also, when we given a subgroup of a permutation group we assume that a base and strong generating set of the subgroup are known.

The main issue is to determine the relationship between the homomorphism ϕ and the stabiliser chains of the domain and image. It is advantageous to choose the bases so that there is a "nice" correspondence between the homomorphism and the stabiliser chains. The first question to ask is:

> What is the preimage of a point stabiliser of the image group?

That is,

> Suppose $\overline{G}$ is the image of the homomorphism ϕ and $\overline{\alpha}$ is a point on which it acts. Which subgroup H of G is the set of all elements which map to the stabiliser $\overline{H} = \overline{G}_{\overline{\alpha}}$?

The correspondence will be "nice" if one can establish a relationship such as

$$\phi\left[G^{(i)}\right] = \phi\left[G\right]^{(i)}$$

between the stabiliser chains and the homomorphism. That is, the correspondence is "nice" if the following diagram commutes.

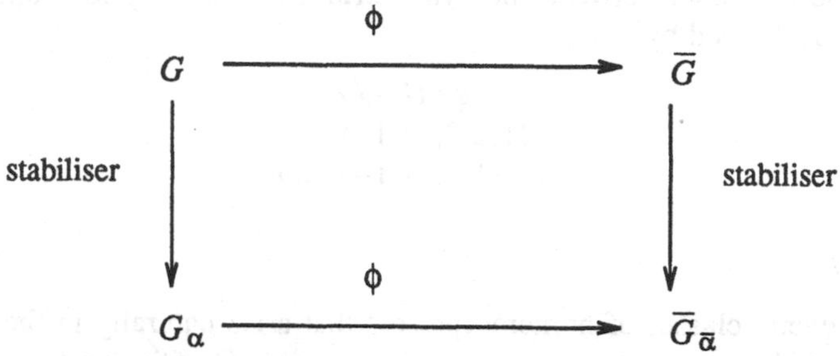

Transitive Constituent Homomorphism

Let the subset Δ of Ω be invariant under the action of G. Let Σ_Δ denote the symmetric group on $\{1,2,\ldots,|\Delta|\}$. Then for each bijection $\overline{}$ between Δ and $\{1,2,\ldots,|\Delta|\}$, there is a homomorphism

$$\phi : G \rightarrow \Sigma_\Delta$$

defined by restricting the action of an element g to Δ (and relabelling). Hence, for $\alpha,\beta \in \Delta$, the image of $\overline{\alpha}$ under $\phi(g)$ is $\overline{\beta}$ if and only if $\alpha^g = \beta$.

Consider the group G to be the symmetries of the square acting on the six pairs of points

1. $\{1,2\}$ 2. $\{1,3\}$ 3. $\{1,4\}$ 4. $\{2,3\}$ 5. $\{2,4\}$ 6. $\{3,4\}$

The action on this set of the two generators are a=*elt*[2]=(1,4,6,3)(2,5) and b=*elt*[5]=(1,3)(4,6). Hence, the two sets $\Delta_1=\{1,3,4,6\}$ and $\Delta_2=\{2,5\}$ are invariant under the action of G. They are the edges and non-edges of the square. For the homomorphism induced by restricting to Δ_1 the relabelling $\overline{}$ can be represented by an vector l

	1	2	3	4	5	6
l	1	0	2	3	0	4

where $l[\alpha]$ is $\overline{\alpha}$, or zero when $\alpha \notin \Delta$. Then if $\alpha \in \Delta$, the entry in the permutation $\overline{g}$ is defined by $\overline{g}[l[\alpha]] = l[\alpha^g]$. Hence, $\overline{a} = (1,3,4,2)$ and $\overline{b} = (1,2)(3,4)$.

For $\alpha \in \Delta$, the stabilizer of $\overline{\alpha}$ in the image of G is simply the image of G_α, since $g \in G$ fixes α if and only if $\phi(g)$ fixes $\overline{\alpha}$. Therefore, if $\alpha_1,\alpha_2,\ldots,\alpha_r$ are points of Δ such that $G_{\alpha_1,\alpha_2,\ldots,\alpha_r}$ is the pointwise stabilizer in G of Δ, then not only is $G_{\alpha_1,\alpha_2,\ldots,\alpha_r}$ the kernel of ϕ, but $[\overline{\alpha}_1,\overline{\alpha}_2,\ldots,\overline{\alpha}_r]$ also forms a base for the image of ϕ. Furthermore, suppose the subset S of G contains a set of generators for each subgroup in the chain

$$G \geq G_{\alpha_1} \geq G_{\alpha_1,\alpha_2} \geq \cdots \geq G_{\alpha_1,\alpha_2,\ldots,\alpha_r},$$

then $\phi(S)$ contains a set of generators of each group in the stabilizer chain of $im\phi$ relative to the base $[\bar{\alpha}_1,\bar{\alpha}_2,\ldots,\bar{\alpha}_r]$. That is, $\phi(S)$ is a strong generating set of $im\phi$ relative to $[\alpha_1,\alpha_2,\ldots,\alpha_r]$.

So, given an arbitrary base and strong generating set of G, we proceed by choosing a base

$$B = [\alpha_1,\alpha_2,\ldots,\alpha_r,\beta_1,\beta_2,\ldots,\beta_s]$$

for G such that

1. $\alpha_1,\alpha_2,\ldots,\alpha_r \in \Delta$, and
2. $G_{\alpha_1,\alpha_2,\ldots,\alpha_r}$ is the pointwise stabilizer of Δ in G.

Applying the base change algorithm if necessary, we may assume a strong generating set S of G relative to B is known.

Let us return to the above example where G is the symmetries of the square acting on pairs of points and Δ is Δ_1, the set of edges of the square. The points $\alpha_1=1$ and $\alpha_2=3$ form a base for G, so $G_{1,3} = G_{1,3,4,6} = <$ identity $>$ (and $s=0$). Hence, $\bar{\alpha}_1=1$ and $\bar{\alpha}_2=2$ form a base for $im\phi$. The stabiliser G_1 is generated by $b{\times}a$=(2,5)(3,4), so $\{a,b,b{\times}a\}$ is a strong generating set of G relative to the chosen base. Furthermore, the stabiliser of 1 in $im\phi$ is generated by $\overline{b{\times}a}$=(2,3). Hence, the set of images $\{\ \bar{a}, \bar{b}, \overline{b{\times}a}\ \} = \{$ (1,3,4,2), (1,2)(3,4), (2,3) $\}$ is a strong generating set of $im\phi$ relative to the base [1,2]. The kernel of the homomorphism is the trivial subgroup, $<$ identity $>$.

We may perform each of the basic tasks as follows:

1. Constructing the image of the group: Choose enough points $\alpha_1, \alpha_2, \ldots, \alpha_r \in \Delta$ such that their pointwise stabiliser fixes all points in Δ. Change the base of G to a base B which has initial segment $[\alpha_1,\alpha_2,\ldots,\alpha_r]$. Let S be the corresponding strong generating set, and let $[\beta_1,\beta_2,\ldots,\beta_s]$ complete the base B. The sequence $\bar{B} = [\bar{\alpha}_1,\bar{\alpha}_2,\ldots,\bar{\alpha}_r]$ is a base for $im\phi$, and the set $\phi(S-S^{(r+1)})$ forms a strong generating set of $im\phi$ relative to $\bar{B}$.

Note that the restriction of an element of $S^{(r+1)}$ is the identity. Furthermore, the set $\phi(\,S-S^{(r+1)}\,)$ may contain redundant strong generators. These may be eliminated by applying Algorithm 4 of Chapter 12.

2. Constructing the kernel: The kernel of ϕ is $G_{\alpha_1,\alpha_2,\ldots,\alpha_r}$ which has $[\beta_1,\beta_2,\ldots,\beta_s]$ as a base and $S^{(r+1)}$ as a strong generating set relative to this base.

3. Constructing the image of an element: The image of an element g of G is formed by restricting its action to Δ and relabelling using the bijection $\bar{\ }: \Delta \rightarrow \{1,2,\ldots,|\Delta|\}$. This bijection is easily stored as a vector l indexed by Ω.

4. Constructing the preimage of an element: Given an element h in the image $im\phi$, we determine points $\gamma_1,\gamma_2,\ldots,\gamma_r$ in Δ such that $\bar{\alpha}_i^h = \bar{\gamma}_i$, $1 \leq i \leq r$, by using the inverse of the relabelling $\bar{\ }$. Then the base B for G and any strong generating set of G relative to B allow the computation of an element g in G which maps α_i to γ_i, $1 \leq i \leq r$. The element g belongs to $\phi^{-1}(h)$. In fact, $\phi^{-1}(h)$ is the right coset $(ker\phi)\,g$.

The inverse of $\bar{\ }$ is readily stored as a vector *l_inv* indexed by $\{1,2,..., |\Delta|\}$. For our running example, the vector *l_inv* is

	1	2	3	4
l_inv	1	3	4	6

Algorithm 1 : Preimage of an Element

Input: a group G acting on Ω;
  an invariant subset Δ of Ω;
  a corresponding base $B = [\alpha_1,\alpha_2,\ldots,\alpha_r,\beta_1,\beta_2,\ldots,\beta_s]$
    and strong generating set of G;
  the vector *l_inv*;
  an element $\bar{g}$ of $im\phi$;
Output: an element $g \in G$ which maps to $\bar{g}$;
begin
  for i := 1 **to** r **do**
    γ_i := *l_inv*[$\bar{\alpha}_i^{\bar{g}}$];
  end for;
  g := an element of G mapping $[\alpha_1,\alpha_2,\ldots,\alpha_r]$ to $[\gamma_1,\gamma_2,\ldots,\gamma_r]$;
end.

For example, the element $\bar{g}$=(1,4)(2,3) requires an element $g \in G$ mapping [1,3] to [*l_inv*[4],*l_inv*[3]]=[6,4]. Tracing the Schreier vectors would determine $g=a^2$=(1,6)(3,4).

5. Constructing the image of a subgroup: Suppose we are given a base and strong generating set of a subgroup H of G. Apply the base change algorithm to H to form a strong generating set T relative to B. Then $\bar{B}$ is a base for $\phi(H)$ and $\phi(T - T^{(r+1)})$ is a strong generating set of $\phi(H)$ relative to $\bar{B}$.

6. Constructing the preimage of a subgroup: Suppose we are given a base and strong generating set of a subgroup $\bar{H}$ of $im\phi$. Apply the base change algorithm to $\bar{H}$ and form a strong generating set $\bar{T}$ of $\bar{H}$ relative to $\bar{B}$. Using the method described in (4), form $T = \{ \phi^{-1}(t) \mid t \in \bar{T} \}$. We claim that $T \cup S^{(r+1)}$ is a strong generating set of $H = \phi^{-1}(\bar{H})$ relative to B.

Clearly B is a base for H, and $S^{(r+1)}$ is a strong generating set of $H^{(r+1)}$. If $T \cup S^{(r+1)}$ is not a strong generating set then there exists a group $H^{(i)}$, $1 \le i \le r$, in the stabilizer chain of H for which the group generated by $T^{(i)} \cup S^{(r+1)}$ is properly contained in $H^{(i)}$. But, since $S^{(r+1)}$ generates $ker\phi$, this implies that $\phi(T^{(i)})$ generates a proper subgroup of $\bar{H}^{(i)}$, thereby contradicting the fact that $\bar{T}$ is a strong generating set of $\bar{H}$ relative to $\bar{B}$.

The algorithms as implemented consist of a major routine which produces a table holding the information necessary to efficiently perform the above tasks. The table stores

a. the vector l describing the bijection $\bar{\ }: \Delta \to \{1,2,..., |\Delta|\}$,

b. a vector *l_inv* for the inverse of $\bar{\ }$,

c. the base $B = [\alpha_1, \alpha_2, \ldots, \alpha_r, \beta_1, \beta_2, \ldots, \beta_s]$,

d. the base $\bar{B} = [\bar{\alpha}_1, \bar{\alpha}_2, \ldots, \bar{\alpha}_r]$ and the strong generating set $\phi(S - S^{(r+1)})$ of $im\phi$, and

e. the base $[\beta_1, \beta_2, \ldots, \beta_s]$ and the strong generating set $S^{(r+1)}$ of $ker\phi$.

Blocks Homomorphism

The homomorphisms of this section were originally defined in terms of systems of imprimitivity, however, all that is required is a partition of the set of points which is invariant under the action of the group. It is not even necessary for the action to be transitive.

A partition $\pi = \{B_1 \mid B_2 \mid \cdots \mid B_t\}$ of Ω is G-invariant if G permutes the subsets $B_1, B_2, \ldots, B_t$ amongst themselves. (The subsets $B_1, B_2, \ldots, B_t$ are also called *blocks*.) Let Σ_t be the symmetric group on $\{1,2,...,t\}$. There is a homomorphism

$$\phi : G \rightarrow \Sigma_t$$

where the image $\phi(g)$ of a permutation g maps i to j if and only if $B_i^g = B_j$.

As an example consider the group G of the symmetries of the square acting on the 4 vertices of the square. Then π={1,3|2,4} is an invariant partition. The partition could be represented by a vector π mapping the points of Ω to an integer in {1,2} indicating whether the point was in the first or second subset of the partition:

	1	2	3	4
π	1	2	1	2

We also need to know a representative point b_i of each block B_i of the partition. A vector b indexed by $\{1,2,..,t\}$ can represent this information:

	1	2
b	1	2

Given an element $g \in G$ the image $\bar{g}$ is computed by setting $\bar{g}[i] = \pi[b[i]^g]$, for each $i \in \{1,2,...,t\}$. So the image of a=(1,2,3,4) is $\bar{a}$=(1,2).

A permutation $\phi(g)$ stabilizes i if and only if g fixes the block B_i as a set. Let G_{B_i} denote the setwise stabilizer of the block B_i. Then the image of G_{B_i} is the stabilizer of i in $im\phi$. Furthermore, the image of a set of generators of G_{B_i} will generate the stabilizer of i in $im\phi$. We therefore choose blocks $B_{i_1}, B_{i_2}, \ldots, B_{i_r}$ in π such that any permutation which stabilizes each of the blocks $B_{i_1}, B_{i_2}, \ldots, B_{i_r}$ stabilizes every block of π. That is,

$$G_{B_{i_1}, B_{i_2}, \ldots, B_{i_r}} = \bigcap_{B \in \pi} G_B = ker\phi.$$

Then $[i_1, i_2, \ldots, i_r]$ is a base for $im\phi$. A strong generating set of $im\phi$ relative to this base may be computed by taking the images of the sets of generators of each group in the chain

$$G \geq G_{B_{i_1}} \geq G_{B_{i_1}, B_{i_2}} \geq \cdots \geq G_{B_{i_1}, B_{i_2}, \ldots, B_{i_r}}.$$

The first problem then is to compute a base and strong generating set of the stabilizer G_B of a block B. We assume that a base and strong generating set of G are known. Once such an algorithm is known, it may be applied repeatedly to form each group in the chain.

Let $B \in \pi$ and let b be any point in the block B. It follows from the definition of G-invariant that if a permutation g maps b to a point in B then g fixes B setwise. That is, $g \in G_B$. An immediate consequence is that $G_b \leq G_B$. Furthermore, G_b is the stabilizer of b in G_B. Another consequence is that the orbit of b under G_B is precisely the intersection of B with the orbit b^G. These facts may be used in conjunction with the following lemma to prove the correctness of the algorithm that we present below.

Lemma

Let T be a subset of a permutation group H. If T contains a strong generating set of H_b, for some point b, and if the orbit of b under H is the same as the orbit of b under $< T >$, then T is a strong generating set of H (relative to some base beginning with b).

A block B is fixed by $G_{B_{i_1}}$ if and only if the orbit of b under $G_{B_{i_1}}$ is contained in B for any point b in B. These facts lead to an algorithm, which we call *blocks_image*, to produce a base and strong generating set of $im\phi$ and also a base and strong generating set of $ker\phi$.

Algorithm 2 : Blocks Image

Input: a base and strong generating set of a group G acting on Ω;
an invariant partition $\pi = \{B_1 \mid B_2 \mid \dots \mid B_t\}$ of Ω;

Output: a base $\bar{B}$ and a strong generating set $\bar{S}$ of $im\phi$;
a base and strong generating set T of K = $ker\phi$, for blocks homomorphism ϕ;

```
begin (* blocks_image *)
  K := G; T := strong generators of G; B̄ := [ ]; S̄ := empty set;
  for i := 1 to t do (* K is G_{B_1,B_2,...,B_{i-1}} *)
    choose a point b in B_i; form Γ := b^K - {b};
    if not ( Γ ⊆ B_i ) then (* non-redundant block to stabilize *)
      change base of K to start with b;
      T := strong generating set of K_b; Γ := Γ ∩ B_i;
      (* extend T to generate K_{B_i} *)
      while Γ ≠ empty set do
        choose γ ∈ Γ;
        take g = u^(1)(γ) of K as an element mapping b to γ;
        T := T ∪ {g}; Γ := Γ − b^<T>;
      end while;
      (* add next group to stabilizer chain of imφ *)
      append i to B̄;
      add { φ(g) | g ∈ T } to S̄;
      K := <T>; (* T is strong generating set of K relative to present base for K *)
    end if;
  end for;
end. (* blocks_image *)
```

Note that the transitivity of G is not required for the algorithm to be correct. It is only necessary that the group G permute the blocks amongst themselves.

Let us return to our example of the group G, the symmetries of the square acting on the 4 vertices, and the invariant partition $\{1,3|2,4\}$. The first block is $\{1,3\}$ with representative point 1. The stabiliser G_1 is generated by b, and the set Γ is $\{3\}$, so we find an element g mapping 1 to 3. Such an element is a^2=(1,3)(2,4), so the block stabiliser $G_{\{1,3\}}$ is $< a^2, b >$. This subgroup also stabilises the other block $\{2,4\}$. Hence, it is the kernel of the homomorphism. The kernel has a base [2] and strong generating set $\{a^2,b\}$. Therefore, the image $im\phi$ has a base [1] and strong generating set $\{(1,2)\} = \{\bar{a}\}$, since $b \in ker\phi$.

The inverse problem arises during the determination of preimages of subgroups. In this situation we know a base and strong generating set of G_B, for some block B, and we know a set of generators of G. The problem is to determine a base and strong generating set of G. Equivalently, we require a base and strong generating set for some point stabilizer G_α in G. The solution is to choose a point b in B as the point α. Then the stabilizer G_b is precisely the stabilizer of b in G_B, which can be determined by applying the base change algorithm to G_B. Inserting the point b at the start of the base for G_b, and adding the generators of G to the strong generating set of G_b gives the desired base and strong generating set of G.

Before presenting methods to perform the six basic tasks, we will relate the blocks homomorphism to a transitive constituent homomorphism.

We may regard G as acting on the set $\pi \cup \Omega$. Then π is invariant under G, and the blocks homomorphism is precisely the transitive constituent homomorphism. We form a base $[B_{i_1}, B_{i_2}, \ldots, B_{i_r}, \beta_1, \beta_2, \ldots, \beta_s]$ where

$$G_{B_{i_1},B_{i_2},\ldots,B_{i_r}} = \bigcap_{B \in \pi} G_B$$

and form the corresponding strong generating set. Here, however, we work with the representation of G on Ω and use the algorithms of this section to convert from block stabilizers to point stabilizers, and vice versa.

1. Constructing the image of the group: Algorithm *blocks_image* computes a base and strong generating set of $im\phi$.

2. Constructing the kernel: Algorithm *blocks_image* computes a base and strong generating set of $ker\phi$.

3. Constructing the image of an element: If we store a list of representative points $b_1, b_2, \ldots, b_t$ for the blocks $B_1, B_2, \ldots, B_t$, and also store a vector indexed by Ω which gives the number of the block containing a given point, then it is straightforward to determine $\phi(g)$ from an element g of G.

4. Constructing the preimage of an element: Given an element h in $im\phi$, its preimage is determined by working in the representation of G on Ω rather than on $\pi \cup \Omega$, which complicates matters slightly. For each group $G_{B_{i_1},B_{i_2},\ldots,B_{i_{m-1}}}$, $1 \le m \le r$, we store a set of generators and the Schreier vector of the orbit of b_{i_m}. In order to find an element mapping B_{i_m} to B_j, we choose a point γ in the intersection of B_j and the orbit of b_{i_m}, and take the element mapping b_{i_m} to γ as determined by the Schreier vector.

Algorithm 3 : Preimage of an Element

Input: a group G acting on Ω;
an invariant partition π of Ω with vectors π and b;
the corresponding blocks homomorphism ϕ;
a corresponding base $\overline{B} = [i_1, i_2, \ldots, i_r]$ of $im\phi$;
generators of $G_{B_{i_1}, B_{i_2}, \ldots, B_{i_{m-1}}}$, $1 \le m \le r$;
the orbit $\Delta^{(m)}$ and Schreier vector $v^{(m)}$ of b_{i_m} under the
subgroup $G_{B_{i_1}, B_{i_2}, \ldots, B_{i_{m-1}}}$, $1 \le m \le r$;
an element $\overline{g}$ of $im\phi$;
Output: an element $g \in G$ which maps to $\overline{g}$;

```
begin
  for m := 1 to r do
    γ_m := b[ i_m^ḡ ];
  end for;

  g := identity of G;

  for m := 1 to r do
    j := π[ γ_m ];
    find a point γ ∈ B_j ∩ Δ^(m);
    u_m := trace( γ, v^(m) );
    g := u_m × g;
    [γ_1, γ_2, ..., γ_r] := [γ_1, γ_2, ..., γ_r]^(u_m^-1);
  end for;
end.
```

5. Constructing the image of a subgroup: Given a base and strong generating set of a subgroup H of G, then algorithm *blocks_image* will determine a base and strong generating set of $\phi(H)$ since the blocks of π are permuted amongst themselves by H.

6. Constructing the preimage of a subgroup: Let $\overline{H}$ be a subgroup of $im\phi$, and let $H = \phi^{-1}(\overline{H})$. As in the case of the transitive constituent homomorphism, we can form generators for each group in the chain

$$H \ge H_{B_{i_1}} \ge H_{B_{i_1}, B_{i_2}} \ge \cdots \ge H_{B_{i_1}, B_{i_2}, \ldots, B_{i_r}}$$

by changing the base of $\overline{H}$ to $[i_1, i_2, \ldots, i_r]$ and using the methods in (4) above. The problem is to use this information to form generators for a chain of point stabilizers in H; that is, form a base and strong generating set of H. Since $H_{B_{i_1}, B_{i_2}, \ldots, B_{i_r}}$ is $ker\phi$ (and we know a base and strong generating set of $ker\phi$ by (2) above), the problem is to work up the chain of block stabilizers forming a base and strong generating set of $H_{B_{i_1}, B_{i_2}, \ldots, B_{i_{m-1}}}$ using a base and strong generating set of $H_{B_{i_1}, B_{i_2}, \ldots, B_{i_m}}$ and the generators of $H_{B_{i_1}, B_{i_2}, \ldots, B_{i_{m-1}}}$. This is precisely the inverse problem solved above.

The complete process is presented as algorithm *blocks_preimage*. The strong generating set of H produced may contain redundancies.

Algorithm 4 : Blocks Preimage

Input: a group G acting on Ω;
an invariant partition $\pi = \{B_1 | B_2 | \ldots | B_t\}$ of Ω;
the corresponding blocks homomorphism ϕ;
a base $\overline{B} = [i_1, i_2, \ldots, i_r]$ of $im\phi$;
a strong generating set $\overline{T}$ of a subgroup $\overline{H}$ of $im\phi$ relative to the base $\overline{B}$;
a base and strong generating set of $ker\phi$;

Output: a base and strong generating set T of $H = \phi^{-1}(\overline{H})$;

begin (* blocks_preimage *)

(* form a base and strong generating set of $H_{B_{i_1}, B_{i_2}, \ldots, B_{i_r}}$ *)
$H := ker\phi$;

(* work up the chain of block stabilizers *)
for $m := r$ **downto** 1 **do**
(* form a base and strong generating set of $H_{B_{i_1}, B_{i_2}, \ldots, B_{i_{m-1}}}$ *)
choose a point b in B_{i_m}; change base of H to start with b;
$T := \phi^{-1}(\overline{T}^{(m)}) \cup$ strong generators of H_b;
$H := \langle T \rangle$; (* with base of previous H and strong generating set T *)
end for;

end. (* blocks_preimage *)

The table of information required by the blocks homomorphism routines is as follows.

a. a list of representative points $b_1, b_2, \ldots, b_t$ of the blocks,

b. a vector giving the partition π of Ω; that is a map from Ω to $\{1,2,\ldots,t\}$,

c. a base $[i_1, i_2, \ldots, i_r]$ and strong generating set of $im\phi$,

d. a base $[\beta_1, \beta_2, \ldots, \beta_s]$ and strong generating set of $ker\phi$, and

e. for each m, $1 \le m \le r$, generators of $G_{B_{i_1}, B_{i_2}, \ldots, B_{i_{m-1}}}$ and the Schreier vector of the orbit of b_{i_m} under the action of this group.

Other Homomorphisms

General homomorphisms may be defined by specifying the image of each generator of the group G. This section looks at such homomorphisms. Let G be a permutation group acting on $\Omega_1 = \{1,2,\ldots,n\}$ and generated by the set S. Let H be a permutation group acting on $\Omega_2 = \{n+1, n+2, \ldots, n+m\}$. Let $f : S \to H$ be a map specifying the image of each generator of G. The map f may determine a homomorphism ϕ defined by

$$\phi : G \to H$$
$$s_1^{\varepsilon_1} \times \cdots \times s_t^{\varepsilon_t} \mapsto f(s_1)^{\varepsilon_1} \times \cdots \times f(s_t)^{\varepsilon_t}$$

The first problem is that not all maps $f : S \to H$ determine a homomorphism. It may be the case that if we express an element g as a word in two different ways - say as w_1 and w_2- then the "image" of g will be different. That is, $f(w_1) \neq f(w_2)$. Then the homomorphism is not even well-defined. In this case we have that the word $w_1 \times w_2^{-1}$ represents the identity element - that is, it is a relator - but it does not get mapped to the identity under f. Actually, f determines a homomorphism if and only if every relator of G is mapped to the identity element of H.

The computations with the map f and the homomorphism ϕ are carried out in the direct product

$$G \times H = \{ (g,h) \mid g \in G, h \in H \}$$

which is the group of all pairs (g,h). The multiplication of these pairs is defined componentwise

$$(g_1,h_1) \times (g_2,h_2) = (g_1 \times g_2, h_1 \times h_2)$$

and so is inversion

$$(g,h)^{-1} = (g^{-1}, h^{-1})$$

The identity element of the direct product is $(identity_G, identity_H)$. The direct product acts on the set $\Omega = \Omega_1 \cup \Omega_2 = \{1,2,...n+m\}$ by defining the image of a point i under (g,h) to be i^g if $1 \leq i \leq n$, and to be i^h if $n+1 \leq i \leq n+m$.

Define the subgroup $F = < (s,f(s)) \mid s \in S >$ of $G \times H$. We can rephrase the criterion that f maps every relator to the identity as

Lemma

The map f determines a homomorphism if and only if

$$F \cap \left[\{ identity_G \} \times H \right] = \{ identity_{G \times H} \}$$

Similarly, the definition of the kernel as the subgroup of G of all the elements which map to the identity can be rephrased as

Lemma

If the map f determines a homomorphism ϕ, then

$$ker\phi = F \cap \left[G \times \{ identity_H \} \right]$$

The definition of a base as being a sequence of points whose stabiliser is the identity allows us to rephrase the above lemmas as

Theorem

Let B be a base for G acting on Ω_1, and let C be a base for H acting on Ω_2. Then

(a) the map f determines a homomorphism if and only if B is a base for F.

(b) if the map f determines a homomorphism ϕ, then $ker\phi$ is the pointwise stabiliser of C in F.

As an example, consider the group G to be the symmetric group of degree 4 acting on $\Omega_1 = \{1,2,3,4\}$ and generated by $a=(1,2,3,4)$ and $b=(1,2,3)$. So $S=\{a,b\}$. Let H be the cyclic group of order 2 acting on the set $\Omega_2 = \{5,6\}$ and generated by $z=(5,6)$. Define $f: S \rightarrow H$ by $a \mapsto z$ and $b \mapsto identity$. Then $G \times H$ acts on $\Omega = \{1,2,3,4,5,6\}$ and is generated by

$$(a,\text{identity})=(1,2,3,4)(5)(6)$$
$$(b,\text{identity})=(1,2,3)(4)(5)(6)$$
$$(\text{identity},z)=(1)(2)(3)(4)(5,6)$$

The subgroup F is generated by

$$(a,z)=(1,2,3,4)(5,6)$$
$$(b,\text{identity})=(1,2,3)(4)(5)(6)$$

A base B for G is [1,2,3] and a base C for H is [5]. Using the Schreier-Sims method, we can verify that B is a base for F, and that $ker\phi$ is $< b,a^2 >$, the stabiliser F_5.

We now turn to how we can perform each of the basic tasks. We will assume we know a base B and strong generating set T for G, that we know a generating set S of G, and that we know a base C for H. The map f is defined on S, though it is useful if f is defined on T.

0. Checking that f determines a homomorphism: Form the subgroup F and using the Schreier-Sims method verify that B is a base for F.

If we already know that f determines a homomorphism ϕ then we can directly form a strong generating set of F relative to B by taking the set $\{ (t,\phi(t)) \mid t \in T \}$, where T is a strong generating set of G relative to B.

In either case, we can assume that we know a base and strong generating set for F after we have confirmed that f determines a homomorphism ϕ.

1. Constructing the image of the group: A base C and strong generating set of $im\phi$ acting on Ω_2 can be constructed by changing the base of F to begin with C and restricting each element of the resulting strong generating set of F to Ω_2.

2. Constructing the kernel: The kernel is the stabiliser of C in F, so changing the base of F to begin with C will determine a base and strong generating set of $ker\phi$ acting on Ω. We can restrict to Ω_1 if we so desire.

3. Constructing the image of an element: Given an element $g \in G$ we take its base image B^g and determine the element of F which maps its base B to B^g. The restriction of this element to Ω_2 is $\phi(g)$.

4. Constructing the preimage of an element: Given an element $h \in im\phi$ we take its base image C^h and determine an element of F which maps C to C^h using a base D for F which has C as an initial segment. The restriction of this element to Ω_1 is $\phi^{-1}(h)$.

5. Constructing the image of a subgroup: Given a subgroup L of G with a strong generating set X relative to B, we form the subgroup L_F of $G \times H$ which has base B and strong generating set $\{ \ (x,\phi(x)) \mid x \in X \ \}$. We change the base of L_F to begin with C and restrict the elements of the resulting strong generating set to Ω_2. The set of these restricted elements is a strong generating set of $\phi(L)$ relative to the base C.

6. Constructing the preimage of a subgroup: Given a subgroup L of $im\phi$ with a strong generating set X relative to C, we form the subgroup L_F of $G \times H$ with a base D (the base for F which begins with C) and a strong generating set which is the union of $\{ \ (x,\phi(x)) \mid x \in X \ \}$ and a strong generating set of the stabiliser of C in F (relative to D). We change the base of L_F to be B and restrict the elements of the resulting strong generating set to Ω_1. The set of these restricted elements is a strong generating set of $\phi^{-1}(L)$ relative to the base B.

Summary

This chapter has considered the tasks of computing images and preimages of elements and subgroups under homomorphisms. In particular, it deals with the task of computing the kernel of the homomorphism. Most attention is given to the natural homomorphisms of a permutation group that arise from invariant subsets and partitions of the points, but the chapter also considers general homomorphisms defined in terms of the images of the group generators.

The algorithms determine a base and strong generating set of the subgroups constructed as kernels, images, or preimages. They are efficient, and the most expensive component in their execution is the base change algorithm.

Homomorphisms play an important role in divide-and-conquer algorithms for permutation groups, as the reader will see in the following chapters.

Exercises

(1/Easy) Consider the group G to be the symmetries of the square acting on the pairs of points. Take the set Δ to be the set Δ_2 of non-edges of the square. Let ϕ be the corresponding transitive constituent homomorphism. Determine an appropriate base of G, a base and strong generating set of $im\phi$, and a base and strong generating set of $ker\phi$.

(2/Moderate) Let G be the symmetries of the square acting on the 6 pairs of points. The partition $\{1,6|2,5|3,4\}$ is invariant under the action of G. Determine the block stabiliser K of $\{1,6\}$, and then the block stabiliser of $\{2,5\}$ in K. Hence, give a base and strong generating set for $im\phi$, where ϕ is the corresponding blocks homomorphism.

(3/Moderate) Consider the sixth group G of Chapter 10. The group has degree 14 and is generated by

$$a=(1,2)(3,4)(5,6)(7,8)(9,10)(11,12)$$
$$b=(1,13)(2,3,7,5)(6,9,11,8)(10,14)$$

The following partition is invariant under G.

$$\{1,10|2,9|3,11|4,12|5,6|7,8|13,14\}$$

Form the successive block stabilisers of $\{1,10\}$, $\{2,9\}$, $\{3,4\}$ and show that $K = G_{\{1,10\},\{2,9\},\{3,4\}}$ is the kernel of the blocks homomorphism ϕ.

Furthermore, show that K has a base [3,2,1,4,7,8] and strong generating set

(1,10)(2,9)(3,11)(5,6)
(1,10)(2,9)(5,6)(7,8)
(1,10)(4,12)(7,8)(13,14)
(4,10)(13,14)
(2,8)(13,14)
(5,6)(13,14)

Hence, K has order 2^6.

Furthermore, show that the block stabilisers are generated as follows:

$$G_{\{1,10\},\{2,9\}} = < K, (1,10)(3,8)(4,13)(5,6)(7,11)(12,14), (3,14)(4,8)(7,12)(11,13) >$$
$$G_{\{1,10\}} = < G_{\{1,10\},\{2,9\}}, (2,4)(3,11)(5,13)(6,14)(7,8)(9,12) >$$

Hence, show that $im\phi$ has a base [1,2,3] and strong generating set

(1,2)(3,4)
(1,7)(2,3,6,5)
(2,4)(5,7)
(3,6)(4,7)
(3,7)(4,6)

(The image is isomorphic to the symmetries of the projective plane of order two.)

(4/Easy) Consider the group G to be the symmetric group of degree 4 acting on $\Omega_1 = \{1,2,3,4\}$ and generated by a=(1,2,3,4) and b=(1,2,3). So $S=\{a,b\}$. Let H be the cyclic group of order 2 acting on the set $\Omega_2 = \{5,6\}$ and generated by z=(5,6). A base B for G is [1,2,3] and a base C for H is [5].

(a) Define $f: S \rightarrow H$ by $a \mapsto z$ and $b \mapsto z$. The subgroup F is generated by

(a,z)=(1,2,3,4)(5,6)
(b,z)=(1,2,3)(4)(5,6)

Using the Schreier-Sims method - or otherwise - verify that B is not a base for F, and that f does not determine a homomorphism. Find a relator of G in terms of a and b which is not mapped to the identity by f.

(b) Let c=(1,2) of G and take the set S of generators of G to be $\{a,c\}$. Define $f: S \rightarrow H$ by $a \mapsto z$ and $c \mapsto z$. The subgroup F is generated by

(a,z)=(1,2,3,4)(5,6)
(c,z)=(1,2)(3)(4)(5,6)

Using the Schreier-Sims method - or otherwise - verify that B is a base for F, and that f determines a homomorphism ϕ. Compute the stabiliser F_5 and determine $ker\phi$.

Bibliographical Remarks

The natural homomorphisms for permutation groups have been used extensively in the theoretical literature on permutations groups. It has long been clear how to determine images of elements, and hence generators for the image of a subgroup. It was also known that the kernel of the transitive constituent homomorphism defined by a subset Δ is the pointwise stabiliser of Δ. With the development of the Schreier-Sims method and the base change algorithm, people knew how to compute (a base and strong generating set for) the kernel of a transitive constituent homomorphism. Algorithms to perform the remaining tasks for the natural homomorphisms of permutation groups were first developed in the author's thesis G. Butler, **Computational Approaches to Certain Problems in the Theory of Finite Groups**, Ph. D. Thesis, University of Sydney, 1980, and published in G. Butler, *"Effective computation with group homomorphisms"*, Journal of Symbolic Computation **1**, 2 (1985) 143-157.

The results of the last section dealing with general homomorphisms defined by the generator images is due to C.R. Leedham-Green, C.E. Praeger, and L.H. Soicher, *"Computing with group homomorphisms"*, to appear in Journal of Symbolic Computation.

Chapter 16. Sylow Subgroups

This chapter presents an algorithm for computing a Sylow p-subgroup of a permutation group. The algorithm uses particular homomorphisms in a divide-and-conquer fashion to reduce the computation to simpler groups.

Homomorphic Images of Centralizers

The methods that rely on homomorphisms to reduce the problem must address the issue of solving the problem (or a related one) in the image and kernel of the homomorphism, and to somehow combine those two solutions. The aim is to choose the homomorphism so that these tasks are easy. Generally, the combination will involve taking the preimage of the solution for the image. The preimage will contain the kernel if we are dealing with subgroups, or the preimage will be a coset of the kernel if we are dealing with an element. Therefore it is important that the kernel be very closely related to the final solution. In the problem of this chapter we wish to compute a Sylow subgroup. This is a subgroup of order p^m, where p^m is the largest power of the prime p dividing the order of G. If $f : G \rightarrow H$ is a homomorphism, and S is a Sylow p-subgroup of $f(G)$, then the preimage $f^{-1}(S)$ always contains a Sylow subgroup of G. The preimage will be a Sylow subgroup if and only if the kernel of f is a p-group.

The first approach we consider uses a homomorphism that has a p-group as its kernel. However, we cannot guarantee that such homomorphisms exist when the domain is G, but we can if we restrict the domain to be a subgroup of G. Of course, we want the subgroup to contain a Sylow p-subgroup of G. The subgroup we use as the domain is a centralizer C in G of an element z of order p.

Let P be a Sylow p-subgroup of G. The centre $Z(P)$ of P is the subgroup of all elements of P which commute with every element of P. As P is a p-group, it has a non-trivial centre, and if z is any element in $Z(P)$ then $C_G(z)$ contains P. We can always take z to have order p (by taking a suitable power of the element). So we can always find a suitable element z and compute its centralizer C. The elements of C permute the cycles of z so the partition π of Ω given by the cycles of z is invariant under C and determine a blocks homomorphism. The elements in the kernel of this homomorphism commute with z and fix each cycle of z. Hence, each element in the kernel is a element of order p, and the kernel is a p-group.

Theorem

Let $z \in C$ be central and of order p. Let π be the partition of Ω given by the cycles of z, and let $C|_\pi$ be the homomorph of C that acts on the cycles. Let

$$f : C \to C|_\pi$$

be the blocks homomorphism that maps g to its action on the cycles. If $S \in Syl_p(C|_\pi)$ then $f^{-1}(S) \in Syl_p(C)$.

A straightforward recursive application of this result leads to Algorithm 1.

Algorithm 1 : Sylow Using Centralizers

function sylow(*G*:group; *p*:prime):group;
(* Return a Sylow *p*-subgroup of the permutation group *G* *)
begin
  if *p* does not divide the order of *G* **then**
    result is <identity>;
  end if;
  find an element z of order p where $C = C_G(z)$ contains a Sylow p-subgroup of G;
  π := cycles of z;
  let $f : C \to C|_\pi$ be the blocks homomorphism;
  result is f^{-1}(sylow($C|_\pi, p$));
end;

We define the Sylow p-subgroup to be the identity subgroup when p does not divide the order of G in order to simplify the statement of Algorithm 1. In such cases, the result S is the kernel of the homomorphism f. There are additional termination criteria for the recursion that help improve efficiency. Suppose the order of the required Sylow subgroup is p^m. If $m = 1$, then the subgroup $<z>$ is a Sylow p-subgroup. If C is a p-group, then C is a Sylow p-subgroup.

As an example, consider the fourth group G of Chapter 10. The group G has degree 21 and is generated by

a=(1,8,9)(2,11,15)(3,10,12)(4,14,19)(5,16,17)(6,21,20)(7,13,18),
b=(9,18,20)(12,19,17), and
c=(10,21,11)(13,16,14).

It has order $27{,}783 = 3^4 \times 7^3$. A base for the group is

[1, 9, 8, 10, 2, 12]

and a strong generating set relative to this base is

$s_1 = a$, $s_2 = b$, $s_3 = c$,
s_4=(8,13,21)(10,14,16), s_5=(2,6,3)(4,5,7), and s_6=(12,20,15)(17,19,18).

To the compute the Sylow 7-subgroup, we could choose the element z = (9,15,12,19,20,17,18), which has a centralizer C of order $3087 = 3^2 \times 7^3$ which is generated by

(10,21,11)(13,16,14), (8,13,21)(10,14,16),
(2,3,6)(4,7,5), (9,18,17,20,19,12,15), and (1,7,6)(3,4,5).

The partition π is

{ 1 | 2 | 3 | 4 | 5 | 6 | 7 | 8 | 9,12,15,17,18,19,20 | 10 | 11 | 13 | 14 | 16 | 21 }

Let f be the corresponding homomorphism. The kernel of f is just $<z>$ of order 7. The image is a group I of degree 15 and order $441 = 3^2 \times 7^2$. I is generated by

(10,15,11)(12,14,13), (2,3,6)(4,7,5),
(1,7,6)(3,4,5), and (8,12,15)(10,13,14).

Applying the algorithm recursively to I, we find a p-central element z_I = (1,2,3,4,6,5,7)(8,11,10,13,15,14,12), whose centralizer in I is a Sylow 7-subgroup of order 7^2. It is generated by

(8,12,14,15,13,10,11), and (1,7,5,6,4,3,2).

The preimage is a Sylow 7-subgroup of G of order 7^3 and it is generated by

(8,11,10,14,21,16,13), (1,2,3,4,6,5,7), and (9,15,12,19,20,17,18).

Restricting to Orbits

Rather than use a single homomorphism and require that the kernel is a p-group, we can use several homomorphisms so that their combined use will remove any non-p part of the final preimage. What we require is two or more homomorphisms such that the intersection of their kernels is the identity. Then successive application of them will eventually result in a trivial kernel.

Theorem

Let $\Omega = \Gamma_1 \cup \Gamma_2$ where Γ_1 and Γ_2 are invariant under the action of G. Let $G \mid_{\Gamma_1}$ be the restriction of G to the set Γ_1. Let f_1 be the natural homomorphism

$$f_1 : G \to G \mid_{\Gamma_1}$$

Let $S_1 \in Syl_p(G \mid_{\Gamma_1})$, let S be $f_1^{-1}(S_1)$, and let f_2 be the natural homomorphism

$$f_2 : S \to S \mid_{\Gamma_2}$$

If $S_2 \in Syl_p(f_2(S))$ then $f_2^{-1}(S_2) \in Syl_p(G)$.

Algorithm 2 : Sylow Using Centralizers and Orbits

function sylow(G:group; p:prime):group;

(* Return a Sylow p-subgroup of the permutation group G *)

begin
  if p does not divide the order of G **then**
    result is <identity>;
  end if;

  find an element z of order p where $C = C_G(z)$ contains a Sylow p-subgroup of G;

  if z is fixed point free **then**
    result is central_sylow(C, z, p);
  else
    Γ_2 := fixed points of z;
    $\Gamma_1 := \Omega - \Gamma_2$;
    let $f_1 : C \to C|_{\Gamma_1}$ be natural homomorphism;
    $S := f_1^{-1}$(central_sylow($f_1(C), f_1(z), p$));
    let $f_2 : S \to S|_{\Gamma_2}$ be natural homomorphism;
    result is f_2^{-1}(sylow($f_2(S), p$));
  end if;
end;

function central_sylow(G:group; z:element; p:prime):group;

(* Given an element z of order p that is fixed point free and central in G, return a Sylow p-subgroup of G *)

begin
  if degree of G is p **then**
    result is <z>;
  else if transitive(G) **then**
    π := partition of Ω determined by the cycles of z;
    let $f : G \to G|_\pi$ be the blocks homomorphism;
    result is f^{-1}(sylow($G|_\pi, p$));
  else
    Γ_1 := nontrivial orbit of G;
    $\Gamma_2 := \Omega - \Gamma_1$;
    let $f_1 : G \to G|_{\Gamma_1}$ be natural homomorphism;
    $S := f_1^{-1}$(central_sylow($f_1(G), f_1(z), p$));
    let $f_2 : S \to S|_{\Gamma_2}$ be natural homomorphism;
    result is f_2^{-1}(central_sylow($f_2(S), f_2(z), p$));
  end if;
end;

The function *central_sylow* requires the element z to be fixed-point-free so that its restrictions $f_1(z)$ and $f_2(z)$ are assured of not being the identity element. Hence, the restrictions are p-central in the quotient, and we can directly call *central_sylow* rather than *sylow*.

The recursion in function *central_sylow* means that we restrict to each nontrivial orbit of G in the case where G is intransitive. This could be done iteratively.

There are some ways to improve efficiency. If the preimage S under f_1 is a p-group - that is, the kernel of f_1 just happens to be a p-group - then S is a Sylow p-subgroup and it is not necessary to form f_2 and compute the Sylow subgroup of its image.

Let us again consider the Sylow 7-subgroup of the example group G. Again we choose the element z = (9,15,12,19,20,17,18), which has centralizer C of order $3087 = 3^2 \times 7^3$ which is generated by

(10,21,11)(13,16,14), (8,13,21)(10,14,16),
(2,3,6)(4,7,5), (9,18,17,20,19,12,15), and (1,7,6)(3,4,5).

The fixed points of z are

$$\Gamma_2 = \{1,2,3,4,5,6,7,8,10,11,13,14,16,21\},$$

and the non-fixed points are

$$\Gamma_1 = \{9,15,12,19,20,17,18\}.$$

The homomorphism f_1 has a kernel of order $441 = 3^2 \times 7^2$ generated by

(10,21,11)(13,16,14), (8,13,21)(10,14,16),
(2,3,6)(4,7,5), and (1,7,6)(3,4,5).

while the image is a cyclic group of order 7 generated by the image of z. Hence, S is the centralizer C.

The homomorphism f_2 of S has a kernel of order 7 generated by z. The image is a group of degree 14 and order $441 = 3^2 \times 7^2$. The image I is generated by

(9,14,10)(11,13,12), (8,11,14)(9,12,13),
(2,3,6)(4,7,5), and (1,7,6)(3,4,5).

Applying the algorithm recursively to I, we find a p-central element z_I = (1,2,3,4,6,5,7)(8,11,13,14,12,9,10), whose centralizer in I is a Sylow 7-subgroup of order 7^2. It is generated by

(8,10,9,12,14,13,11), and (1,7,5,6,4,3,2).

The preimage is a Sylow 7-subgroup of G of order 7^3 and generated by

(8,11,10,14,21,16,13), (1,2,3,4,6,5,7), and (9,15,12,19,20,17,18).

Finding Centralizers

So far we have ignored the problem of locating a p-central element in G. In practice, an element of order p is sought by examining the orders of the terms in a random sequence of elements until one is found having order divisible by p. By taking a suitable power of such an element, one obtains an element x of order p. By computing the centralizer $C_G(x)$, we can determine whether or not x is p-central.

In the case of simple and near-simple groups G, such a random search will often locate a p-central element after generating two or three random elements. However, for other classes of groups, such as soluble groups, such a random method may have to examine an impractically large number of elements in order to have any chance of locating a p-central element.

In such cases we may resort to the following algorithm:

Algorithm 3 : Locating p-Central Elements

```
function p_central( G:group; p:prime ):element;
(* Given a group G of order p^m s, where m>0 and p does not divide s,
return a p-central element of G *)
begin
    x := a randomly chosen element of order p;
    C := C_G(x);
    let p^r be the largest power of p dividing |C|;
    while r ≠ m do
        P := Sylow p-subgroup of C;
        x := an element of Z(P) of order p such that p^(r+1) divides |C_G(x)|;
        C := C_G(x);
        let p^r be the largest power of p dividing |C|;
    end while;
    result is x;
end;
```

Consider the group G of order $27{,}783 = 3^4 \times 7^3$ for the prime p=3. There are 4256 elements of order p, of which only 686 are p-central. There are 12,348 elements which power to a p-central element, and 10,837 elements which power to an element of order 3 which is not p-central. There are only 343 elements whose order is not divisible by 3. Hence, the chances that a random element will power to a p-central element is approximately 50% and all is well for this group.

If, however, we did follow Algorithm 3 and chose a non-p-central element x of order 3, then the centralizer C has a Sylow 3-subgroup P of order 3^3 which is elementary abelian. Hence, the centre of P is the whole of P. Of the 26 elements in $Z(P)$ only 2 are p-central, so the algorithm may need to consider 25 elements before locating a p-central one.

Random Elements

In determining random elements of the group, we wish to obtain each element with equal probability. For a group G with a base and strong generating set this can be done by independently choosing a random coset representative $u_i \in U^{(i)}$, for each i, and taking the random element $u_k \times u_{k-1} \times \cdots \times u_1$. Selecting a random coset representative requires randomly choosing an integer in the range 1 to $|U^{(i)}|$ with uniform distribution, or equivalently choosing a random point in the set $\Delta^{(i)}$.

For a group G given only by a set of generators S the situation is difficult, and we cannot guarantee a uniform distribution of random elements. One approach used is to consider a random word w in the generators S of a given length and then evaluating the word. Each symbol in w is chosen randomly from the set S (and perhaps the set of inverses). It is a simple enough task to choose each symbol of w randomly with uniform distribution. However, there are two major problems:

1. To ensure that it is possible to generate each element of the group in this way requires excessively long words: for example, the dihedral group of degree n and order $2 \times n$ is generated by two elements - $s_1 = (1,n)(2,n-1)(3,n-2)...(n/2,n/2+1)$ and $s_2 = (1,n-1)(2,n-2)(3,n-3)...(n/2-1,n/2+1)$ - and the element $(s_1 \times s_2)^{n/2}$ can not be generated by a word of length less than n.

2. Even if we can generate each element of G by words of the specified length there is no guarantee that the distribution is uniform.

The long words required are also expensive to evaluate.

An approach used to reduce this expense is to only use long words as seeds to the random generator. Each seed w is evaluated to an element g and used to generate a small number of elements by randomly choosing a generator s by which to multiply the current element. After this short sequence of random elements has been generated, a new seed is chosen.

Algorithm 4 : Random Elements

Input: a group G given by a set of generators S;

Output: a sequence of pseudo-random elements of G;

```
begin
    determine l, the length of words required, from |S| and |Ω|;
    while true do
        choose a random word w of length l;
        evaluate w to obtain g;
        for i := 1 to small_number do
            choose a random generator s ∈ S;
            g := g × s;   (* the next random element *)
        end for;
    end while;
end;
```

More on Blocks Homomorphisms

We have previously used a block homomorphism which was guaranteed to have a p-group as a kernel. However, we can follow the lead of the transitive constituent homomorphisms and choose two homomorphisms whose kernels have trivial intersection. If two block systems are distinct and minimal then the corresponding blocks homomorphisms will have trivial intersection, and we can utilise the following result.

Theorem

Let G be transitive on Ω and let ρ_1 and ρ_2 be distinct minimal systems of imprimitivity of G. Let $G|_{\rho_1}$ be the action of G induced on the subsets of ρ_1. Let f_1 be the natural homomorphism

$$f_1 : G \rightarrow G|_{\rho_1}$$

Let $S_1 \in Syl_p(G|_{\rho_1})$, let S be $f_1^{-1}(S_1)$, and let f_2 be the natural homomorphism

$$f_2 : S \rightarrow S|_{\rho_2}$$

If $S_2 \in Syl_p(f_2(S))$ then $f_2^{-1}(S_2) \in Syl_p(G)$.

There is some possibility to reduce the problem in the case where G has only one minimal block system. It just may be the case that the kernel is a p-group, or the preimage of a Sylow p-subgroup of its image may be a proper subgroup of G. Provided that G itself is not a p-group, these techniques will fail to reduce the problem precisely when the image is a p-group. Other techniques (like those of Algorithm 1) must then be used.

If the group is primitive then there are no block systems to use. If the degree of G is not divisible by p, then the point stabiliser contains a Sylow p-subgroup, and we can reduce the problem.

Algorithm 5 : Sylow Using Orbits and Blocks

```
function sylow( G:group; p:prime ):group;

(* Return a Sylow p-subgroup of the permutation group G *)

begin
  if p does not divide the order of G then
    result is <identity>;
  end if;

  if not transitive(G) then
    for each orbit Γ of G do
      let f : S → S |_Γ be natural homomorphism;
      S := f^-1( sylow( f(S), p ) );
      if S is a p-group then break; end if;
    end for;
    result is S;
  else
    result is transitive_sylow( S, p );
  end if;
end;

function transitive_sylow( G:group; p:prime ):group;

(* Return a Sylow p-subgroup of a transitive permutation group G *)

begin
  if degree of G is not divisible by p then
    result is sylow( G_1, p );
  else if G has minimal block systems ρ1, ρ2 then
    let f1 : G → G |_ρ1 be natural homomorphism;
    S := f1^-1( sylow( f1(G), p ) );
    let f2 : S → S |_ρ2 be natural homomorphism;
    result is f2^-1( sylow( f2(S), p ) );
  else if G has minimal block system ρ1 then
    let f1 : G → G |_ρ1 be natural homomorphism;
    S := f1^-1( sylow( f1(G), p ) );
    if S = G then
      (* algorithm fails *)
    else
      result is sylow( S, p );
    end if;
  else (* G is primitive and p divides degree *)
    (* algorithm fails *)
  end if;
end;
```

As an example, again consider the fourth group G of Chapter 10. The group G has degree 21 and is generated by

a=(1,8,9)(2,11,15)(3,10,12)(4,14,19)(5,16,17)(6,21,20)(7,13,18),
b=(9,18,20)(12,19,17), and
c=(10,21,11)(13,16,14).

It has order 27,783=$3^4 \times 7^3$. The group is imprimitive with one minimal block system ρ given by

{ 1,2,3,4,5,6,7 | 8,10,11,13,14,16,21 | 9,12,15,17,18,19,20 }.

Let f be the corresponding homomorphism. The kernel of f is a subgroup K of order 9261 = $3^3 \times 7^3$ generated by

(9,18,20)(12,19,17), (10,21,11)(13,16,14),
(12,20,15)(17,19,18), (1,7,5,6,4,3,2),
(8,13,21)(10,14,16), (1,7,3)(2,4,6), and (2,3,6)(4,7,5).

The image is a cyclic group of order 3, so by convention it has trivial Sylow 7-subgroup, and the preimage S is just K.

The algorithm is applied recursively to K. The group is intransitive with three orbits of length 7 corresponding to the three blocks of ρ. Let f_1 be the homomorphism of K corresponding to the orbit {1,2,3,4,5,6,7}. The kernel of f_1 has order 441 = $3^2 \times 7^2$ and is generated by

(9,18,20)(12,19,17), (10,21,11)(13,16,14),
(12,20,15)(17,19,18), and (8,13,21)(10,14,16).

The image is a group of order 21 = 3×7 which has a Sylow 7-subgroup generated by (1,2,3,4,6,5,7). The preimage of this Sylow 7-subgroup is S of order $3^2 \times 7^3$ generated by

(1,2,3,4,6,5,7), (9,18,20)(12,19,17), (8,13,21)(10,14,16),
(10,11,21)(13,14,16), and (12,15,20)(17,18,19).

Let f_2 be the homomorphism of S corresponding to the orbit {8,10,11,13,14,16,21}. The kernel of f_2 has order 147 = 3×7^2 and is generated by

(1,2,3,4,6,5,7),
(9,18,20)(12,19,17), and (12,15,20)(17,18,19).

The image is a group of order 21 = 3×7 which has a Sylow 7-subgroup generated by (1,2,7,4,3,5,6). The preimage of this Sylow 7-subgroup is S of order 3×7^3 generated by

(8,14,13,10,16,11,21), (1,2,3,4,6,5,7),
(9,18,20)(12,19,17), and (12,15,20)(17,18,19).

Let f_3 be the homomorphism of S corresponding to the orbit {9,12,15,17,18,19,20}. The kernel of f_3 has order 49 = 7^2 and is generated by

(8,14,13,10,16,11,21), and (1,2,3,4,6,5,7).

The image is a group of order $21 = 3 \times 7$ which has a Sylow 7-subgroup generated by (1,2,7,5,3,6,4). The preimage of this Sylow 7-subgroup is S of order 7^3 generated by

(9,19,18,12,17,15,20), (8,14,13,10,16,11,21), and (1,2,3,4,6,5,7).

Primitive Groups

All is not lost if the group G is primitive. The primitive permutation groups have been classified by what is called the O'Nan-Scott Theorem into very well-specified cases. However, to date only some of the cases have practical solutions to reducing the problem of computing Sylow p-subgroups, and we will not pursue them in this chapter.

Summary

This chapter has demonstrated how homomorphisms can be used to reduce the problem of computing a Sylow p-subgroup to smaller cases. This divide-and-conquer approach to analysing permutation groups and solving problems is very general and powerful. We will see further examples in later chapters.

The reductions mentioned in this chapter can be combined with each other and combined with the cyclic extension method or methods for treating soluble permutation groups. The best mix is a matter of engineering and is still the focus of further investigation.

Exercises

(1/Easy) Exercise 1(i) of Chapter 11 asks you to compute the centralizer of z=(1,2)(4,7) in the symmetries G of the projective plane of order 2. This element is p-central, for $p = 2$. Use the centralizer to compute a Sylow 2-subgroup of G.

(2/Easy) Exercise 1(ii) of Chapter 11 asks you to compute the normalizer of $<$ (1,2)(4,7), (4,7)(5,6) $>$ in the symmetries G of the projective plane of order 2. Restrict the normalizer to its action on the orbit of length 3 and hence compute a Sylow 2-subgroup of G.

(3/Easy) Find a 2-central element of the symmetric group of degree 4, compute its centralizer, and compute a Sylow 2-subgroup.

(4/Moderate) Let G be the symmetries of the projective plane of order 2. Execute Algorithm 5 for this example and the prime $p = 2$. As p does not divide the degree 7, we work within G_1 and restrict G_1 to its orbit of length 6. The image has one block system and the kernel of the corresponding homomorphism is a group of order 4. Take a random element of order 2 in the image of the blocks homomorphism and obtain a Sylow 2-subgroup of this image. Hence, compute a Sylow 2-subgroup of G.

(5/Moderate) Exercise 3 of Chapter 15 deals with a group G of degree 14 and order $2^9 \times 3 \times 7$, which is the sixth group of Chapter 10. G has one block system and the corresponding homomorphism has a kernel of order 2^6. Use the information in Exercise 3 of Chapter 15 and compute a Sylow 2-subgroup of G.

Bibliographical Remarks

The problem of constructing a Sylow subgroup for a moderately large group was first considered in J.J. Cannon, *"Computing local structure of large finite groups"*, **Computers in Algebra and Number Theory,** G. Birkhoff and M. Hall, Jr (eds), (SIAM-AMS Proceedings, Vol. 4), American Mathematical Society, Providence, R.I., 1971, pp. 161-176. The Sylow algorithm presented in that paper works for general groups. It is based on the cyclic extension technique (of chapter 6). It begins by finding a p-element, and searches centralizers for the extending elements. The algorithm can find Sylow subgroups in groups of order up to one million.

This algorithm was implemented for permutation groups by the author in 1977 to specifically use the stabiliser chain of the centralizer during a backtrack search for extending elements and to find a strong generating set. The work is described in the author's Ph.D. thesis of 1979 and in G. Butler and J.J. Cannon, *"Computing in permutation and matrix groups III: Sylow subgroups"*, Journal of Symbolic Computation, **8** (1989) 241-252. This backtrack method for computing Sylow subgroups begins to struggle on permutation groups of degrees in the hundreds, or those groups whose order is divisible by a high power of p, say p^{10} or higher. The techniques described in this chapter using homomorphisms were introduced to overcome these limitations. The use of homomorphic images of centralizers is first studied in G. Butler and J.J. Cannon, *"Using homomorphisms to compute Sylow subgroups of permutation groups"*, TR 222, Basser Department of Computer Science, University of Sydney, 1984, along with limited use of restricting to orbits in order to reduce the degree of the group. The bottleneck of finding a p-element z was particularly evident in soluble groups, so further development of this approach had to wait for the work of S.P. Glasby, *"Computing normalisers in finite soluble groups"*, Journal of Symbolic Computation, **5** (1988) 285-294, on computing a Sylow subgroup of a soluble group given by an AG-system and the work (in 1987 and 1989 respectively) of C.C. Sims, *"Computing the order of a solvable permutation group"*, Journal of Symbolic Computation, **9** (1990) 699-705, and G. Butler, *"Computing a conditioned pc presentation of a soluble permutation group"*, TR 392, Basser Department of Computer Science, University of Sydney, 1990, to set up the isomorphism between a soluble permutation group and an appropriate AG-system (see chapter 18 for more details). The resulting algorithm is presented in G. Butler and J.J. Cannon, *"Computing Sylow subgroups of permutation groups via homomorphic images of centralizers"*, to appear in Journal of Symbolic Computation.

Around 1982, Derek Holt at Warwick implemented an algorithm which searches the normalizer of a p-group for the extending elements required by the cyclic extension method. It works up the stabiliser chain of the group G and begins with the first level of index divisible by p. A heuristic looks at random elements from the current level of the chain for one that can be powered to a p-element which extends the p-group. If the heuristic fails then the normalizer is computed and searched for one extending element. The process is repeated until a Sylow subgroup of the stabiliser is found, and then the algorithm proceeds to the next higher level whose index is divisible by p. This work is hinted at in D.F. Holt, *"A computer program for the calculation of the Schur multiplier of a permutation group"*, **Computational Group Theory,** M.D. Atkinson (ed.), Academic Press, Academic Press, 1984, pp. 307-319, and D.F. Holt, *"Computing normalizers in permutation groups"*, to appear in Journal of Symbolic Computation.

More use of the natural homomorphisms of permutation groups was made by M.D. Atkinson and P.M. Neumann, *"Computing Sylow subgroups of permutation groups"*, (Twenty-first Southeastern Conference on Combinatorics, Graph Theory, and Computing, Boca Raton, February 1990), to appear in Congressus Numeratium. In particular, they exploit the fact a group may have two minimal block systems, and the fact that a primitive group may have an elementary abelian regular normal p-subgroup. In several cases, they restrict to the point stabiliser. In one case, the Sylow subgroup of the point stabiliser is used as the starting point in the cyclic extension method. In that case, they extend the p-group using the techniques of Holt described above. However, there are cases where their approach cannot proceed and must fall back on an alternative way of computing Sylow subgroups.

The O'Nan-Scott Theorem and its use in computation with permutation groups is explained in P.M. Neumann, *"Some algorithms for computing with permutation groups"*, **Groups - St Andrews 1985**, E.F. Robertson and C.M. Campbell (eds), London Mathematics Society Lecture Notes **121**, Cambridge University Press, Cambridge, 1986, pp. 59-92.

There have also been several theoretical investigations of the complexity of the problem. These rely on the classification of simple groups, reductions using homomorphisms, and the recognition of simple groups to give polynomial-time algorithms for computing Sylow subgroups of permutation groups. In the general case, the complexity is $O(|\Omega|^9)$. W.M. Kantor, *"Polynomial-time algorithms for finding elements of prime order and Sylow subgroups"*, Journal of Algorithms, **6** (1985) 478-514, restricts to the case where the group G is simple, while W.M. Kantor and D.E. Taylor, *"Polynomial-time versions of Sylow's theorem"*, Journal of Algorithms, **9** (1988) 1-17, restrict to a group G which is either soluble or has all of its noncyclic composition factors suitably restricted. The general case is treated in W.M. Kantor, *"Sylow's theorem in polynomial-time"*, Journal of Computer Systems and Science, **30** (1985) 359-394. Glasby's algorithm for computing the Sylow subgroup of a soluble group defined by an AG-system is based on Kantor's ideas.

Knowledge of the Sylow p-subgroup of G for various primes p dividing the order of G is crucial when analyzing the structure of G. In particular, the construction of a Sylow p-subgroup is a basic step in current algorithms for computing such things as the maximal normal p-subgroup of G, $O_p(G)$; the Fitting subgroup; the socle of G; representatives of conjugacy classes of elements of prime power order; and the first and second cohomology groups of G.

Chapter 17. P-Groups and Soluble Groups

A p-group G is a group of order p^n, for some prime p and integer n. A soluble group is a group whose derived series reaches the identity subgroup. Both p-groups and soluble groups can be described by pc presentations, and can be studied using algorithms which exploit the properties of the series of subgroups associated with the pc presentation. Both classes of groups arise in special cases of advanced algorithms for permutation groups, and often those special cases are best dealt with by computing a pc presentation and then utilising an algorithm exploiting the pc presentation. This chapter briefly describes computation with pc presentations for p-groups and soluble groups. We restrict our attention to finite groups.

Presentations

A *presentation* of a group G in terms of generators and relations consists of a set X (of generators) and a set R of relations $u=v$, where u, v are words in X. For example, the presentation

$$< a, b \mid a^2 = b^3 = (ab)^4 = 1 >$$

of S_4 specifies the generators a, b and the relations $a^2=1$, $b^3=1$, and $(ab)^4=1$, where 1 represents the empty word (which corresponds to the identity element of the group). The words can include the inverses of the generators.

The group described by the presentation is isomorphic to a quotient of the *free* group F generated by X. The elements of F are equivalence classes of the words in X. Two words are equivalent if and only if they *freely reduce* to the same word. Free reduction repeatedly replaces all subwords of the form $x^{-1}x$ and xx^{-1} by the empty word until no such subwords exist. Hence, $bab^{-1}aa^{-1}ba$ freely reduces to baa. So the only identification of words in F is determined by the group axiom: $g \times g^{-1} = g^{-1} \times g =$ identity, for all elements g. Multiplication in F is performed by concatenating a word from each equivalence class (and then possibly freely reducing the word to obtain a class representative). The inverse of a word $x_1^{k_1} x_2^{k_2} \cdots x_n^{k_n}$ is the word $x_n^{-k_n} x_{n-1}^{-k_{n-1}} \cdots x_1^{-k_1}$.

The relations of the presentation of a group G specify further identifications of words of F. The relation $u=v$ says that u and v represent the same element, or that uv^{-1} represents the identity element. In terms of the Cayley graph of G, the word uv^{-1} represents a loop (about every node). However, the consequences or deductions of the relations must also be taken into account. These are all the loops in the Cayley graph that can be constructed from the loops given by the relations. These are precisely those obtained by tracing a sequence of loops (corresponding to group multiplication), tracing a loop in the reverse direction (corresponding to inversion), and tracing a word w to some node g then tracing a loop l about g and then tracing w^{-1} back to the starting node (corresponding to the conjugate wlw^{-1}). Let N be the normal closure of R; that is, the subgroup of F generated by the relators uv^{-1}, where $u=v$ is in R, and all consequences of these relators. Then N is a normal subgroup of F since it is closed under conjugation. The group G described by the presentation is the quotient F/N.

PC Presentations of Soluble Groups

A *power-commutation presentation* (pc presentation) of a finite soluble group G has the form

$$< a_1, a_2, ..., a_n \mid a_i^{\rho(i)} = u_i, \quad 1 \le i \le n, \quad a_j a_i = a_i v_{ij}, \quad 1 \le i < j \le n >,$$

where each $\rho(i)$ is a positive integer, and u_i, v_{ij} are words in $\{a_{i+1}, a_{i+2}, ..., a_n\}$. (These are also called *AG systems* or *power-conjugate presentations.*) For example, the symmetric group S_4 has a pc presentation

$$< a_1,a_2,a_3,a_4 \mid a_1^2 = a_3,\ a_2^3 = a_3^2 = a_4^2 = 1,\ a_2a_1 = a_1a_2^2a_3,\ a_3a_1 = a_1a_3,\ a_4a_1 = a_1a_3a_4,\ a_3a_2 = a_2a_4,\ a_4a_2 = a_2a_3a_4,\ a_4a_3 = a_3a_4 >,$$

where a_1 to a_4 correspond to the permutations (1,2,3,4), (1,2,3), (1,3)(2,4), and (1,2)(3,4) respectively.

The group

$$D_{12} = < a_1, a_2 \mid a_1^2 = a_2^6 = 1,\ a_2a_1 = a_1a_2^5 >$$

is the dihedral group of order 12 where a_2 corresponds to (1,2,3,4,5,6) and a_1 corresponds to (2,6)(3,5).

A pc presentation defines a chain of subgroups

$$G = G(1) \ge G(2) \ge \cdots \ge G(n) \ge G(n+1) = < identity >,$$

where $G(i) = < a_i, a_{i+1}, ..., a_n >$. The relations $a_ja_i = a_iv_{ij}$ are equivalent to $a_i^{-1}a_ja_i = v_{ij}$. Since $v_{ij} \in G(i+1)$, the subgroup $G(i+1)$ is a *normal* subgroup of $G(i)$. We call such a chain of subgroups a *subnormal series* of G. The relation $a_i^{\rho(i)} = u_i$ and $u_i \in G(i+1)$ says that the quotient group $G(i)/G(i+1)$ is generated by $a_iG(i+1)$ and that the order of the quotient divides $\rho(i)$. (In most cases, we want the order to be precisely $\rho(i)$ - such pc presentations are said to be *consistent* - however, for an arbitrary pc presentation we can only say that the order divides $\rho(i)$.)

If each $\rho(i)$ is prime, then we say the series is *prime-step*. We can always refine a pc presentation to a prime-step series by introducing new generators which correspond to powers of the generators a_i where $\rho(i)$ is composite. For example, for D_{12} we introduce a_3 corresponding to a_2^2 and obtain the pc presentation

$$< a_1, a_2, a_3 \mid a_1^2 = 1,\ a_2^2 = a_3,\ a_3^3 = 1,\ a_2a_1 = a_1a_2^5,\ a_3a_1 = a_1a_3^2,\ a_3a_2 = a_2a_3 >$$

or instead of $a_1a_2^5$ we could have written $a_1a_2a_3^2$.

The relations of a pc presentation allow any word in the generators to be reduced to a *normal word*

$$a_1^{k_1}a_2^{k_2} \cdots a_n^{k_n}, \text{ where } 0 \le k_i < \rho(i),$$

where the generators come in order and the exponents lie in a restricted range.

- The relation $a_i^{\rho(i)} = u_i$ says that a subword $a_i^{n_i}$, where n_i is positive, can be written as $a_i^{k_i}(u_i)^m$ where $n_i = m\rho(i)+k_i$ and $0 \le k_i < \rho(i)$.

- The relation $a^{\rho(i)} = u_i$ says that a subword a_i^{-1} can be written as $a_i^{\rho(i)-1}u_i^{-1}$ since multiplying on the left by a_i gives the identity.
- The relation $a_j a_i = a_i v_{ij}$ for $i < j$ says that a subword $a_j a_i$ where the generators are not in order can be rewritten as $a_i v_{ij}$ where a_i is in order relative to all the generators in v_{ij}.

If the presentation is consistent, then the normal words will be unique; that is, every word in $\{a_1, a_2, ..., a_n\}$ reduces to only one normal word. Then we say the normal words are *canonical*. The order of the group is then $\rho(1)\times\rho(2)\times\cdots\times\rho(n)$, the number of normal words.

The *derived subgroup* (or commutator subgroup) of a group G is the subgroup generated by all commutators $[a,b]$ $(= a^{-1}b^{-1}ab)$ where $a,b \in G$. It is written $[G,G]$ or $D(G)$. It is the smallest normal subgroup of G such that the quotient of G by the normal subgroup is abelian. The *derived series* of G is defined by

a. $D^0(G) = G$, and

b. $D^{c+1}(G) = [D^c(G), D^c(G)]$, for each $c \geq 0$.

A group G is defined to be *soluble* if and only if some term of the derived series of G is the identity subgroup. The derived series is a subnormal series since $D(G)$ is normal in G. It is also a *normal series* since each term $D^c(G)$ is normal in G. Furthermore, each quotient $D^c(G)/D^{c+1}(G)$ is abelian. Each abelian group H can be written as a direct product of cyclic groups of prime power order. Hence, H has a normal series

$$H = H(1) \geq H(2) \geq \cdots \geq H(m) \geq H(m+1) = <\mathit{identity}>,$$

where the quotients $H(i)/H(i+1)$ are *elementary abelian*. Therefore, the derived series of G can be refined to a normal series with elementary abelian quotients, and this *normal* series can be further refined to a prime-step *subnormal* series. This leads to the following definition:

A *conditioned* pc presentation of a finite soluble group G is a consistent pc presentation of the form

$$< a_1, a_2, ..., a_n \mid a_i^{p(i)} = u_i,\ 1 \leq i \leq n,\quad a_j a_i = a_i v_{ij},\ 1 \leq i < j \leq n >,$$

where

1. $p(i)$ is a prime,
2. u_i, v_{ij} are normal words in $\{a_{i+1}, a_{i+2}, ..., a_n\}$, and
3. there exists a *normal* series

   $$G = N_1 \geq N_2 \geq \cdots \geq N_r \geq N_{r+1} = <\mathit{identity}>,$$

   of G where each quotient N_i/N_{i+1} is elementary abelian, and each N_i has the form $< a_{n(i)}, a_{n(i)+1}, \cdots, a_n >$ for some integer $n(i)$.

Every finite soluble group has a conditioned pc presentation, and the extra properties are very useful when computing in a finite soluble group.

The example for S_4 is conditioned where $N_1 = G$, $N_2 = < a_2, a_3, a_4 >$, $N_3 = < a_3, a_4 >$, and $N_4 = <\mathit{identity}>$. The prime-step pc presentation for D_{12} with the relation $a_2 a_1 = a_1 a_2 a_3^2$ is conditioned where $N_1 = G$, $N_2 = < a_2, a_3 >$, $N_3 = < a_3 >$, and $N_4 = <\mathit{identity}>$.

PC Presentations of p-Groups

A *power-commutator presentation* (pc presentation) of a finite p-group G is a presentation of the form

$$< a_1, a_2, ..., a_n \mid a_i^{\rho(i)} = u_i, \ 1 \le i \le n,$$
$$[a_j, a_i] = v_{ij}^*, \ 1 \le i < j \le n >,$$

where each $\rho(i)$ is a power of the prime p, u_i are words in $\{a_{i+1}, a_{i+2}, ..., a_n\}$, and v_{ij}^*, are normal words in $\{a_{j+1}, a_{j+2}, ..., a_n\}$.

The relations $[a_j, a_i] = v_{ij}^*$ can be rewritten as $a_j a_i = a_i a_j v_{ij}^*$ and allow words to be reduced to normal words (as was the case for pc presentations of soluble groups). The pc presentation is *consistent* if every word reduces to a unique normal word. The order of the group G is then $\rho(1) \times \rho(2) \times \cdots \times \rho(n)$. If the pc presentation is prime-step - that is, each $\rho(i) = p$ - then $|G| = p^n$.

The pc presentations of p-groups are related to descending central chains of subgroups. A chain

$$G_0 \ge G_1 \ge \cdots \ge G_c \ge G_{c+1} \ge \cdots$$

of subgroups of a group G is a *descending central chain* of subgroups of G if

a. $G_0 = G$, and

b. $G_{c+1} \ge [G_c, G]$, for each c.

The *lower central chain* of G is defined by $\gamma_0(G) = \gamma_1(G) = G$ and $\gamma_{c+1}(G) = [\gamma_c(G), G]$ for $c \ge 1$. A group is *nilpotent* if some $\gamma_{c+1}(G) = \{\text{identity}\}$. All p-groups are nilpotent.

The *lower exponent-p-central* chain for a prime p is

$$G = P_0(G) \ge P_1(G) \ge \cdots \ge P_c(G) \ge P_{c+1}(G) \ge \cdots$$

where $P_{c+1}(G) \ge [P_c(G), G] \left[P_c(G)\right]^p$, for each c.

(Warning: There is much inconsistency in the literature about whether series begin at the zero-th term or the first term.)

Some important properties of these chains are summarised now.

1. Every descending central series is a *subnormal* series; that is G_{c+1} is normal in G_c.
2. Every descending central series is a *normal* series; that is G_{c+1} is normal in G.
3. Every descending central series is a *central* series; that is G_c/G_{c+1} is contained in the centre of G/G_{c+1}.
4. Each of the quotients G_c/G_{c+1} of a descending central series is abelian.
5. Each of the quotients $P_c(G)/P_{c+1}(G)$ of the lower exponent-p-central series is an elementary abelian p-group.

A pc presentation of a finite p-group G is *conditioned* if it is prime-step and defines a refinement of the lower exponent-p-central series. That is, it has the form

$$< a_1, a_2, ..., a_n \mid a_i^p = u_i,\ 1 \le i \le n,\quad [a_j,a_i] = v^*_{ij},\ 1 \le i < j \le n >$$

where u_i are words in $\{a_{i+1}, a_{i+2}, ..., a_n\}$, and v^*_{ij} are normal words in $\{a_{j+1}, a_{j+2}, ..., a_n\}$. Furthermore, there are integers $w(c)$ such that

$$P_c(G) = < a_{w(c)}, a_{w(c)+1}, \cdots, a_n >$$

for each c.

The *weight* of a generator a_i (denoted $wt(a_i)$) is the value c such that $w(c) \le i < w(c+1)$. The generators of weight 0 are the *defining generators*. There are d ($= w(1)-1$) of them.

For example, the following pcp for the quaternion group of order 8:

$$< a_1, a_2, a_3 \mid a_1^2 = a_3,\ a_2^2 = a_3,\ a_3^2 = 1,\quad [a_2,a_1] = a_3,\ [a_3,a_1] = 1,\ [a_3,a_2] = 1 > \qquad \text{(Q8)}$$

is conditioned with $w(0) = 1$, $w(1) = 3$ and $w(2) = 4$. So $d = 2$.

For example, the group of order 2^6 given by the conditioned pcp

$$< b_1, b_2, b_3, b_4, b_5, b_6 \mid b_2^2 = b_4,\ b_3^2 = b_5,\ b_4^2 = b_5,\quad [b_2,b_1] = b_3,\ [b_3,b_1] = b_5,\ [b_3,b_2] = b_6,\ [b_4,b_1] = b_5 b_6 > \qquad \text{(G64)}$$

- it is usual to omit powers and commutators which are trivial - has $d = 2$, $w(0) = 1$, $w(1) = 3$, $w(2) = 5$, and $w(3) = 7$.

The defining generators are important because any algorithm which needs to compute an action by G only needs to use the action of the defining generators and not the action of all n generators.

The algorithms for p-groups that we will consider in this chapter do not use the weights of the generators or the fact that the presentation is conditioned. However, such presentations are necessary for many algorithms.

Collection

A *collection process* is an algorithm for determining an equivalent normal form of a word in $\{a_1, a_2, ..., a_n\}$. The algorithms look for subwords which prevent the word being normal and rewrite the subword. Table 1 gives (A) a list of (minimal) subwords which violate the property of being normal, and (B) the corresponding rewrite of the subword.

(A)	(B) soluble group	p-group	
a_i^{-1}	$a_i^{\rho(i)-1} u_i^{-1}$	$a_i^{\rho(i)-1} u_i^{-1}$	$1 \le i \le n$
$a_i^{\rho(i)}$	u_i	u_i	$1 \le i \le n$
$a_j a_i$	$a_i v_{ij}$	$a_i a_j v^*_{ij}$	$1 \le i < j \le n$

The outline of the algorithm is

Algorithm 1 : Outline of collection

Input: a pc presentation with generators $\{a_1, a_2, ..., a_n\}$;
a word w in $\{a_1, a_2, ..., a_n\}$;

Output: a normal word w equivalent to the input word;

begin
while w is not normal **do**
$w :=$ the result of replacing a subword of w which occurs in list (A) by its equivalent in list (B) ;
end while;
end.

As an example consider the pcp of the quaternion group Q8. We will collect the word $a_3{}^{-1}a_2a_1a_2a_1{}^{-1}$ and highlight in bold the subword we are replacing at each step.

$a_3{}^{-1}a_2a_1a_2\mathbf{a_1}^{-1}$
$\mathbf{a_3}^{-1}a_2a_1a_2a_1a_3{}^{-1}$
$a_3\mathbf{a_2a_1}a_2a_1a_3{}^{-1}$
$a_3a_1a_2a_3\mathbf{a_2a_1}a_3{}^{-1}$
$\mathbf{a_3a_1}a_2a_3a_1a_2a_3a_3{}^{-1}$
$a_1\mathbf{a_3a_2}a_3a_1a_2a_3a_3{}^{-1}$
$a_1a_2\mathbf{a_3}^{2}a_1a_2a_3a_3{}^{-1}$
$a_1\mathbf{a_2a_1}a_2a_3a_3{}^{-1}$
$a_1{}^2a_2\mathbf{a_3a_2}a_3a_3{}^{-1}$
$a_1{}^2a_2{}^2\mathbf{a_3}^{2}a_3{}^{-1}$
$a_1{}^2\mathbf{a_2}^{2}a_3{}^{-1}$
$a_1{}^2a_3\mathbf{a_3}^{-1}$
$a_1{}^2\mathbf{a_3}^{2}$
$\mathbf{a_1}^{2}$
a_3

The collection processes always terminate. To show this we order the words and show that each iteration of the collection process results in a word that is "more normal" than its predecessor. The ordering $\gg$ is called the *collected ordering* (or syllable ordering). First, order the generators and their inverses so that

$$a_1{}^{-1} \gg a_1 \gg a_2{}^{-1} \gg a_2 \gg \cdots \gg a_n{}^{-1} \gg a_n.$$

Then specify that all nonempty words are greater than the empty word. If u is a non-empty word, define $y(u)$ to be the largest generator (or inverse) y in u. Write u as

$$u = A_0yA_1y \cdots yA_r$$

where the generator y does not occur in the subwords A_i. We write $A_i(u)$ and $r(u)$ to denote A_i and r when it is important to specify the word u. Then we define $u \gg v$ for non-empty words u, v if and only if either

1. $y(u) \gg y(v)$,
2. $y(u) = y(v)$ and $r(u) > r(v)$,
3. $y(u) = y(v)$, $r(u) = r(v)$, and for some j, $0 \le j \le r$, $A_j(u) \gg A_j(v)$ and $A_i(u) = A_i(v)$ for $i = 0,1,2...,j-1$.

The ordering $\gg$ is a well-ordering and is *translation invariant*. That is, if $u \gg v$ then $wu \gg wv$ and $uw \gg vw$ for all words w. This implies that u is greater than any subword of u.

From the definition of $\gg$ we can see that each word in list (A) is greater than the corresponding entry in list (B). Since $\gg$ is translation invariant, this means that the result of one replacement is less than its input word. So each iteration of the collection process decreases the word w (with respect to $\gg$) and makes it "more normal".

An efficient collection process is important. Efficiency depends on the strategy followed in choosing which subword to replace, and upon good implementations. The implementation might preprocess the pc presentation if one expects to perform many collections. The main strategies that have been studied are

- collection *from the right* in which the subword that is replaced is always the rightmost one;
- collection *to the left* in which the leftmost occurrence of the greatest (with respect to $\gg$) uncollected generator (or inverse) y is moved to the left by the choice of a suitable subword; (A generator is "uncollected" if it is in a subword which occurs in list (A).)
- collection *from the left* in which the subword that is replaced is always the leftmost one.

The treatment of inverses during the collection process is often simplified by first "clearing inverses" - that is, recursively applying the first row of Table 1 to eliminate inverses - before following one of the above strategies. We will not follow that practice in our examples, but as a consequence we sometimes require a substep to clear an inverse before the collection strategy can be applied.

We will repeat the collection of the word $a_3^{-1}a_2a_1a_2a_1^{-1}$ in Q8 with each of these strategies and highlight in bold the subword we are replacing at each step.

Collection from the right:

$a_3^{-1}a_2a_1a_2\mathbf{a_1^{-1}}$
$a_3^{-1}a_2a_1a_2a_1\mathbf{a_3^{-1}}$
$a_3^{-1}a_2a_1\mathbf{a_2a_1}a_3$
$a_3^{-1}a_2a_1a_1a_2\mathbf{a_3a_3}$
$a_3^{-1}a_2\mathbf{a_1a_1}a_2$
$a_3^{-1}a_2\mathbf{a_3a_2}$
$a_3^{-1}\mathbf{a_2a_2}a_3$
$a_3^{-1}\mathbf{a_3a_3}$
$\mathbf{a_3^{-1}}$
a_3

Collection to the left:

$a_3^{-1}a_2a_1a_2\mathbf{a}_1^{-1}$

$a_3^{-1}\mathbf{a}_2\mathbf{a}_1a_2a_1a_3^{-1}$

$\mathbf{a}_3^{-1}\mathbf{a}_1a_2a_3a_2a_1a_3^{-1}$

... in two substeps ... first a_3^{-1} replaced by a_3

$a_1a_3a_2a_3\mathbf{a}_2\mathbf{a}_1a_3^{-1}$

$a_1a_3a_2\mathbf{a}_3\mathbf{a}_1a_2a_3a_3^{-1}$

$a_1a_3\mathbf{a}_2\mathbf{a}_1a_3a_2a_3a_3^{-1}$

$a_1\mathbf{a}_3\mathbf{a}_1a_2a_3a_3a_2a_3a_3^{-1}$

$\mathbf{a}_1\mathbf{a}_1a_3a_2a_3a_3a_2a_3a_3^{-1}$

$a_3\mathbf{a}_3\mathbf{a}_2a_3a_3a_2a_3a_3^{-1}$

$\mathbf{a}_3\mathbf{a}_2a_3a_3a_3a_2a_3a_3^{-1}$

$a_2a_3a_3a_3\mathbf{a}_3\mathbf{a}_2a_3a_3^{-1}$

$a_2a_3a_3\mathbf{a}_3\mathbf{a}_2a_3a_3a_3^{-1}$

$a_2a_3\mathbf{a}_3\mathbf{a}_2a_3a_3a_3a_3^{-1}$

$a_2\mathbf{a}_3\mathbf{a}_2a_3a_3a_3a_3a_3^{-1}$

$\mathbf{a}_2\mathbf{a}_2a_3a_3a_3a_3a_3a_3^{-1}$

$a_3a_3a_3a_3a_3a_3\mathbf{a}_3^{-1}$

$\mathbf{a}_3\mathbf{a}_3a_3a_3a_3a_3a_3$

$\mathbf{a}_3\mathbf{a}_3a_3a_3a_3$

$\mathbf{a}_3\mathbf{a}_3a_3$

a_3

Collection from the left:

$\mathbf{a}_3^{-1}a_2a_1a_2a_1^{-1}$

$\mathbf{a}_3\mathbf{a}_2a_1a_2a_1^{-1}$

$a_2\mathbf{a}_3\mathbf{a}_1a_2a_1^{-1}$

$\mathbf{a}_2\mathbf{a}_1a_3a_2a_1^{-1}$

$a_1a_2\mathbf{a}_3\mathbf{a}_3a_2a_1^{-1}$

$a_1\mathbf{a}_2\mathbf{a}_2a_1^{-1}$

$a_1a_3\mathbf{a}_1^{-1}$

$a_1a_3a_1\mathbf{a}_3^{-1}$

$a_1\mathbf{a}_3\mathbf{a}_1a_3$

$\mathbf{a}_1\mathbf{a}_1a_3a_3$

$\mathbf{a}_3\mathbf{a}_3a_3$

a_3

The strategy should not only minimise the number of replacements but also control the length of the intermediate words. Current empirical and theoretical evidence strongly suggests that collection from the left is best in general.

Consistency

In the terminology of rewriting systems, a consistent pcp is an example of a *complete* system of rewrite rules. A rewrite rule is a pair (L,R) of words, and a rewrite system is a set of rewrite rules. A system defines a rewriting process on words where an occurrence of L as a subword may be replaced by R until no more replacements are possible. The result of rewriting is called a normal word. A rewrite system is *complete* if the process of rewriting always terminates, and for each word it returns a unique answer irrespective of the order of replacements of R for L. One checks completeness by guaranteeing termination of the rewriting process, and by checking that each *critical pair* rewrites to the same normal word. A critical pair is a minimal example of a word w which can be rewritten in more than one way (together with the pair of rewrite rules (L_1, R_1) and (L_2, R_2) such that L_1 and L_2 are subwords of w).

The rewrite rules for a pc presentation are the pairs (L,R), where L is an entry of list (A) and R is the corresponding entry of list (B). Collection is the rewriting process. For pcp's termination of rewriting is automatic. The critical pairs are determined by considering the minimal words which display an overlap of the subwords in list (A). For a finite soluble group, these overlaps are

(C) - finite soluble group	
$a_k a_j a_i$	$1 \le i < j < k \le n$
$a_j^{p(j)} a_i$	$1 \le i < j \le n$
$a_j a_i^{p(j)}$	$1 \le i < j \le n$
$a_i^{p(j)+1}$	$1 \le i \le n$

Each of these critical pairs has precisely two subwords in list (A). Each subword should be rewritten using the entry in list (B) and then collected to a normal word. It is not important which strategy is used to complete the collection. However, each of the two possible ways of beginning the collection process must be tried separately. The two resulting normal words are equal if the pcp is consistent.

For a finite p-group with a prime-step pc presentation, the overlaps are

(C) - finite p-group	
$a_k a_j a_i$	$1 \le i < j < k \le n$
$a_j^p a_i$	$1 \le i < j \le n$
$a_j a_i^p$	$1 \le i < j \le n$
a_i^{p+1}	$1 \le i \le n$

For cases where we know $a_1, a_2, \ldots, a_d$ are the defining generators of a p-group, the list of overlaps to check can be reduced to

(C') - finite p-group	
$a_k a_j a_i$	$1 \le i < j < k \le n$ and $i \le d$
$a_j^p a_i$	$1 \le i < j \le n$ and $i \le d$
$a_j a_i^p$	$1 \le i < j \le n$
a_i^{p+1}	$1 \le i \le n$

The proof that this reduced list suffices is difficult (and not explained here).

Elements and Subgroups of p-Groups

For ease of exposition, this section restricts discussion to the context of finite p-groups described by a consistent pcp. The algorithms and techniques carry over to finite soluble groups with only minor changes.

The canonical form of an element as a normal word

$$\prod_{i=1}^{n} a_i^{\varepsilon(i)} \quad \text{with} \quad 0 \leq \varepsilon(i) < p.$$

means that we can establish a mapping

$$\exp : G \rightarrow V(n,p)$$

between G and the n-dimensional vector space $V(n,p)$ over Z_p by defining

$$\prod_{i=1}^{n} a_i^{\varepsilon(i)} \quad \mapsto (\varepsilon(1), \varepsilon(2), \ldots, \varepsilon(n)).$$

So the elements of the group can be represented as vectors. For example, in the quaternion group Q8 an element would be represented in a computer by an array

$\varepsilon(1)$	$\varepsilon(2)$	$\varepsilon(3)$

where $\varepsilon(1), \varepsilon(2), \varepsilon(3)$ take values 0 or 1.

The operations on elements are performed using word operations and collection.

- Multiply g and h by collecting $a_1^{\varepsilon(1)(g)} a_2^{\varepsilon(2)(g)} \cdots a_n^{\varepsilon(n)(g)} a_1^{\varepsilon(1)(h)} a_2^{\varepsilon(2)(h)} \cdots a_n^{\varepsilon(n)(h)}$;
- Invert g by collecting $a_n^{-\varepsilon(n)(g)} a_{n-1}^{-\varepsilon(n-1)(g)} \cdots a_1^{-\varepsilon(1)(g)}$;
- Test g and h for equality by comparing $(\varepsilon(1)(g), \varepsilon(2)(g), \ldots, \varepsilon(n)(g))$ and $(\varepsilon(1)(h), \varepsilon(2)(h), \ldots, \varepsilon(n)(h))$;
- Calculate the order $|g|$ of an element g as follows

```
order := 1;
for i := 1 to n do
  if ε(i)(g) ≠ 0 then
    order := order × p;     (* for soluble case use p(i) *)
    g := g^p;
  end if;
end for;
```

Define the *leading coefficient* of an element g to be the first non-zero entry of its corresponding vector, and define the *leading index* to be the index of the leading coefficient in the vector. Denote these by $lc(g)$ and $li(g)$ respectively.

A subgroup H of G has an induced series defined by

$$H(c) = H \cap G(c) = H \cap \langle a_c, a_{c+1}, ..., a_n \rangle$$

where there may be duplicate terms in the series. We can eliminate these terms and determine a canonical set of generators for a subgroup H as follows.

Definition: A sequence $[h_1, h_2, ..., h_r]$ of elements of H is called a *canonical generating sequence* (cgs) relative to a fixed pcp of G if and only if

i. the sequence $h_1, h_2, ..., h_r$ of elements defines a prime-step subnormal series of H;

ii. $li(h_i) > li(h_j)$ for $i > j$;

iii. $lc(h_i) = 1$ for $i = 1, 2, ..., r$;

iv. $\exp(h_j)[li(h_i)] = 0$ for $i \neq j$.

(If only conditions (i) and (ii) hold, then the sequence is called a *generating sequence* (gs).)

This definition requires that the exponent vectors $\exp(h_i)$ form a row-echelonized matrix of a form like

$$\begin{bmatrix} \exp(h_1) \\ \exp(h_2) \\ \exp(h_3) \end{bmatrix} = \begin{bmatrix} 0\,1\,*\,*\,0\,*\,*\,0\,* \\ 0\,0\,0\,0\,1\,*\,*\,0\,* \\ 0\,0\,0\,0\,0\,0\,0\,1\,* \end{bmatrix}$$

Let $[h_1, h_2, ..., h_r]$ be a cgs of H. Define $H_i = \langle h_i, h_{i+1}, ..., h_r \rangle$ for $1 \leq i \leq r+1$. Then

1. for each $h \in H$ there exists a unique h_i with $li(h) = li(h_i)$; Furthermore, $h \in H_i$.
2. if $h, k \in H$ and $li(h) = li(k) = li(h_i)$ then $hH_{i+1} = kH_{i+1}$ if and only if $lc(h) = lc(k)$;
3. the cgs of H is uniquely determined with respect to a fixed pcp of G;
4. the order of H is p^r.

If the sequence is a gs then (1), (2) and (4) still hold.

The cgs can be formed by applying non-commutative Gaussian elimination to the vectors of an arbitrary generating set of H. The algorithm is similar to row-echelonizing a matrix over a finite field, however, our subgroup is closed under p-th powers and commutators of the generating set.

Algorithm 2 : Non-commutative Gaussian Elimination

Input: a set W of elements of G;

Output: a cgs $\overline{h} = [h_1, h_2, ..., h_r]$ of $H = <W>$;

```
procedure reduce_and_insert( g : element; var h̄ : sequence of elements );
(* h̄ = [h_1, h_2, ..., h_r] has the property that li(h_i) < li(h_j) for i < j.
  The element g is echelonized against h̄ and its reduced form
  (possibly) inserted in the sequence. *)

begin (* reduce_and_insert *)
  for i = 1 to r do
    if li(g) = li(h_i) then
      g := g h_i^(-exp(g)[li(h_i)]); (* now li(g) > li(h_i) *)
    else
      make a gap in the sequence between h_{i-1} and h_i for new i-th generator;
      insert g as the new i-th generator;
    end;
  end;
  if g ≠ identity then append g to h̄; end;
end; (* reduce_and_insert *)

begin
  h̄ := empty sequence;
  for each g ∈ W do
    reduce_and_insert( g, h̄ );
    for each h, k in h̄ do
      reduce_and_insert( h^p, h̄ ); reduce_and_insert( [h,k], h̄ );
    end;
  end;

  for each h in h̄ with lc(h) ≠ 1 do
    replace h by h^m where lc(h^m) = 1;
  end;

  for each h, k in h̄ with li(k) > li(h) and exp(h)[li(k)] ≠ 0 do
    replace h by h k^(-exp(h)[li(k)]);
  end;
end.
```

Let us consider the example of $H = <b_2b_4b_5, b_4b_6>$ in the group (G64). The first thing the algorithm does is to insert the first generator $b_2b_4b_5$ as the first entry h_1 of the cgs. We have the matrix

$$\begin{bmatrix} 010110 \end{bmatrix}$$

The algorithm then considers $h_1^2 = b_4 b_5$ and inserts it as h_2.

$$\begin{bmatrix} 010110 \\ 000110 \end{bmatrix}$$

The commutator $[h_1, h_2]$ is the identity, but $h_2^2 = b_5$, which is inserted as h_3.

$$\begin{bmatrix} 010110 \\ 000110 \\ 000010 \end{bmatrix}$$

The commutators $[h_1, h_3]$ and $[h_2, h_3]$ are the identity, as is h_3^2.

The algorithm now considers the second generator $b_4 b_6$. When inserting this element it is reduced against h_2 by forming $b_4 b_6 h_2 = b_6$. This is inserted as h_4.

$$\begin{bmatrix} 010110 \\ 000110 \\ 000010 \\ 000001 \end{bmatrix}$$

All commutators with h_4 are trivial, as is h_4^2, so we have an echelonized form of the subgroup generators satisfying conditions (1) and (2). We need to enforce conditions (3) and (4) to get the final form.

$$\begin{bmatrix} 010000 \\ 000100 \\ 000010 \\ 000001 \end{bmatrix}$$

The subgroup has order 2^4 and cgs $[b_2, b_4, b_5, b_6]$.

This linear algebra view of subgroups and straightforward application of non-commutative Gaussian elimination allow us to easily

- test membership of an element $g \in G$ in a subgroup H of G;
- form a cgs for the subgroup $\langle H, g \rangle$ where $g \in G$;
- form a cgs for the subgroup $\langle H, K \rangle$ where H and K are subgroups of G;
- test containment $K \leq H$ of subgroups of G;
- form the normal closure of a subgroup;
- form a commutator subgroup $[H,K]$ of subgroups of G, and hence various series of G such as the derived series and lower central series.

The reduction of an element $g \in G$ against a cgs of a subgroup H as in the procedure reduce_and_insert determines a canonical (left) coset representative of gH. If $g \in H$ it also determines the normal word for g with respect to the cgs. Thus a cgs for H determines a pcp for H, and if H is normal it determines a pcp for G/H.

Conjugacy Classes of Elements of p-Groups

Two elements h and k of G are *conjugate* if there exists and element $g \in G$ such that $g^{-1}hg = k$. This defines an equivalence relation and the equivalence classes are called *conjugacy classes*. The centralizer of an element h is defined as

$$C_G(h) = \{ c \mid c \in G, c^{-1}hc = h \}.$$

The elements in the conjugacy class of h are in one-to-one correspondence with the cosets $gC_G(h)$. The determination of the conjugacy classes and the corresponding centralizers is often an important precursor to other algorithms.

The algorithm which computes the conjugacy classes works down the chain of factor groups $G/G(i)$. The step of extending the class information of $G/G(i)$ to that of $G/G(i+1)$ relies on the fact that $G(i)/G(i+1)$ is central in $G/G(i+1)$. For each conjugacy class we store the information

- a representative h of the class, and
- a gs $[c_1, c_2, ..., c_m]$ for the centralizer $C_G(h)$.

Note that groups of order p or p^2 are abelian, so each element lies in a conjugacy class of its own, and its centralizer is the whole group. So in considering the inductive step from $G/G(i)$ to $G/G(i+1)$ we will assume the following information is given

1. a p-group $L = G/G(i)$ of order $p^{i-1} \geq p^3$,
2. a generator a_iG_{i+1} for the minimal normal subgroup $N = G(i)/G(i+1)$ of L, and
3. for each conjugacy class of the factor group L/N a representative and a gs of its centralizer in L/N.

The following result presents the justification for the inductive step.

Proposition : Let L be a p-group of order $\geq p^3$. Let N be a minimal normal subgroup of L and h an arbitrary element of L. Let K be the union of conjugates of the coset hN under L/N. Moreover, let $c_1, c_2, ..., c_{m+1} \in L$ such that $[c_1N, c_2N, ..., c_mN]$ is a gs of the centralizer $C_{L/N}(hN)$ and that $<c_{m+1}> = N$.

Then one of the following cases applies.

I. If $[h,c_j] = 1$ for $1 \leq j \leq m$, then K splits into p conjugacy classes under L with representatives $h, hc_{m+1}, ..., hc_{m+1}^{p-1}$, respectively. All these representatives have the same centralizer, and the elements $[c_1, c_2, ..., c_{m+1}]$ form a gs for it.

II. If $[h,c_j] \neq 1$ for some j, $1 \leq j \leq m$, then K consists of just the conjugates of h in L. Choose the largest such value of j. The for each l, $1 \leq l < j$, there is an integer $k(l)$ with $0 \leq k(l) < p$ such that $[h,c_l] = [h,c_j]^{k(l)}$. Let

$$c_l^* = \begin{cases} c_l c_j^{-k(l)} & \text{for } 1 \leq l < j, \\ c_{l+1} & \text{for } j \leq l \leq m. \end{cases}$$

Then the elements $c_1^*, c_2^*, ..., c_m^*$ form a gs for the centralizer $C_L(h)$. □

For example with the quaternion group of order 8 we begin by noting that $G/G(3)$ is abelian of order p^2 and has classes

	class representative	centralizer gs
(1)	$1G(3)$	$a_1G(3), a_2G(3)$
(2)	$a_1G(3)$	$a_1G(3), a_2G(3)$
(3)	$a_2G(3)$	$a_1G(3), a_2G(3)$
(4)	$a_1a_2G(3)$	$a_1G(3), a_2G(3)$

The classes of G are determined by one inductive step. For (1) we have case I, so we get two classes with representatives 1 and a_3 both with a_1, a_2, a_3 as the gs for their centralizer. For (2) we have case II with $j = 2$ and get one class with representative a_1 and gs a_1, a_3 for its centralizer. For (3) we have case II with $j = 1$ and get one class with representative a_2 and gs a_2, a_3 for its centralizer. For (4) we have case II with $j = 2$ and get one class with representative a_1a_2 and gs $a_1a_2a_3, a_3$ for its centralizer. The final result is

class representative	centralizer gs
1	a_1, a_2, a_3
a_3	a_1, a_2, a_3
a_1	a_1, a_3
a_2	a_2, a_3
a_1a_2	$a_1a_2a_3, a_3$

The algorithm is very efficient - quite capable of handling groups of order 2^{50}. A variation of the algorithm computes the centralizer of a given element.

Sylow Subgroups of Soluble Groups

Let G be a finite soluble group described by a conditioned pc presentation with generators $a_1, a_2, ..., a_n$, subnormal series

$$G = G(1) \geq G(2) \geq \cdots \geq G(n) \geq G(n+1) = < identity >,$$

where $G(i) = < a_i, a_{i+1}, ..., a_n >$, and normal series

$$G = N_1 \geq N_2 \geq \cdots \geq N_r \geq N_{r+1} = < identity >,$$

with elementary abelian quotients where $N_i = < a_{n(i)}, a_{n(i)+1}, \cdots, a_n >$ for some integer $n(i)$. Let $p(i) = |G(i):G(i+1)|$ be prime.

Let p be a prime. The aim of this section is to present an algorithm to compute a Sylow p-subgroup S of the soluble group G. The algorithm recursively works down the normal series by computing a Sylow subgroup of G/N_2, G/N_3,..., G/N_r, G; and works up the subnormal series when extending a Sylow subgroup of G/N_i to one for G/N_{i+1}. Working up the subnormal series, the algorithm uses the cyclic extension technique: it finds an element $g \in G(i)-G(i+1)$ (when $p = p(i)$) which normalizes a Sylow subgroup S of $G(i+1)$ and such that $g^p \in S$. The element g is in the coset $a_iG(i+1)$; however, a_i itself while it does normalize $G(i+1)$ may not normalize S. Hence, the algorithm finds an element y which conjugates S^{a_i} to S and then chooses g to be a suitable power of a_iy. The element y exists as any two Sylow p-subgroups of $G(i+1)$ are conjugate. The element a_iy normalizes S.

Algorithm 3 : Sylow Subgroup of a Soluble Group

```
function pc_sylow( G : group; p : prime ) : subgroup;
(* Given a finite soluble group G with a conditioned pc presentation,
and a prime p, return a Sylow p-subgroup of G *)
begin (* pc_sylow *)
    if p does not divide |G| then
        result is <identity>;
    else
        let S/N_r := pc_sylow( G/N_r, p ) and let [h_1, h_2, ..., h_t] be a cgs for S;
        if N_r is a p-group then
            result is S;
        else
            M := pc_sylow( <h_2, h_3, ..., h_t>, p );
            if p_li(h_1) ≠ p then
                result is M;
            else
                y := conj_elt( G(li(h_1)+1), M^(h_1), M, p );
                let q be integer coprime to p such that |h_1 y| = p^m q;
                result is < (h_1 y)^q, M >;
            end if;
        end if;
    end if;
end; (* pc_sylow *)

function conj_elt( G, H, K : group; p : prime ) : element;
(* Given a finite soluble group G with a conditioned pc presentation,
a prime p, and Sylow p-subgroups H = <h_1, h_2, ..., h_t>, K = <k_1, k_2, ..., k_t>
of G such that H/N_r = K/N_r, return an element g ∈ N_r such that H^g = K *)
begin (* conj_elt *)
    if (N_r is a p-group) or (H/N_r = N_r/N_r) then
        result is identity;
    else
        j := li(k_1)+1; L := <k_2, k_3, ..., k_t>; (* = K ∩ G(j) *)
        y := conj_elt( G(j), H ∩ G(j), L, p );
        h := h_1^y;
        if h^-1 k_1 is a p-element then
            result is y;
        else
            let |h^-1 k_1| = p^m q, where q is coprime to p;
            find integers a, b such that a p^m ≡ 1 mod q and b(-p) ≡ 1 mod q;
            x := (h^-1 k_1)^(a p^m);     z := [ x^h (x^2)^(h^2) ... (x^(p-1))^(h^(p-1)) ]^b ;
            result is yz;
        end if;
    end if;
end; (* conj_elt *)
```

The element z in *conj_elt* is required to conjugate h to an element of the coset k_1L, so that it conjugates $\langle h, L\rangle$ to $\langle k_1, L\rangle$. Hence, we want an element $l \in L$ such that $z^{-1}hz = k_1 l$, or equivalently $[h, z] = h^{-1}z^{-1}hz = h^{-1}k_1 l$. The map $z \mapsto [h,z]$ is a linear transformation of N_r regarded as a vector space. The formula in *conj_elt* for z provides a solution to the equation $[h,z] \in h^{-1}k_1L$ (though we will not prove this fact).

As an example, consider the fourth group G of Chapter 10. The group G has degree 21 and is generated by

a=(1,8,9)(2,11,15)(3,10,12)(4,14,19)(5,16,17)(6,21,20)(7,13,18),
b=(9,18,20)(12,19,17), and
c=(10,21,11)(13,16,14).

It has order $27{,}783 = 3^4 \times 7^3$. It is soluble. The following elements of G:

a_1=a=(1,8,9)(2,11,15)(3,10,12)(4,14,19)(5,16,17)(6,21,20)(7,13,18),
a_2=(2,3,6)(4,7,5),
a_3=(8,13,21)(10,14,16),
a_4=b=(9,18,20)(12,19,17),
a_5=(1,2,3,4,6,5,7),
a_6=(8,14,13,10,16,11,21), and
a_7=(9,19,18,12,17,15,20).

define an isomorphism between G and the group described by the pcp

$$\begin{aligned}
< a_1, a_2, a_3, a_4, a_5, a_6, a_7 \mid\ & a_1^3 = a_2^3 = a_3^3 = a_4^3 = a_5^7 = a_6^7 = a_7^7 = 1,\\
& a_2a_1 = a_1a_3a_6^5,\ a_3a_1 = a_1a_4,\ a_4a_1 = a_1a_2a_5^6,\\
& a_5a_1 = a_1a_6^5,\ a_6a_1 = a_1a_7,\ a_7a_1 = a_1a_5^3,\\
& a_3a_2 = a_2a_3,\ a_4a_2 = a_2a_4,\ a_5a_2 = a_2a_5^2,\ a_6a_2 = a_2a_6,\ a_7a_2 = a_2a_7,\\
& a_4a_3 = a_3a_4,\ a_5a_3 = a_3a_5,\ a_6a_3 = a_3a_6^2,\ a_7a_3 = a_3a_7,\\
& a_5a_4 = a_4a_5,\ a_6a_4 = a_4a_6,\ a_7a_4 = a_4a_7^2,\\
& a_6a_5 = a_5a_6,\ a_7a_5 = a_5a_7,\ a_7a_6 = a_6a_7\ >.
\end{aligned}$$

This is a conditioned pcp with normal series $N_1 = G$, $N_2 = \langle a_2,a_3,\ldots,a_7\rangle$, $N_3 = \langle a_5,a_6,a_7\rangle$, and $N_4 = \langle \mathit{identity}\rangle$. N_3 is an elementary abelian group of order 7^3; N_2/N_3 is an elementary abelian group of order 3^3, and N_1/N_2 is a cyclic group of order 3.

For $p = 7$ the computation first calculates a Sylow 7-subgroup S/N_3 of G/N_3. The result is $S = N_3$ since G/N_3 has order 3^4. Then it returns the result N_3 as a Sylow 7-subgroup of G since N_3 is a 7-group.

For $p = 3$, the computation is more involved. Since G/N_3 is a 3-group, the call *pc_sylow*(G/N_3, 3) returns G/N_3, so $S = G$. N_3 is not a 3-group, so we calculate *pc_sylow*($\langle a_2,a_3,\ldots,a_7\rangle$, 3), which is $M = \langle a_2,a_3,a_4\rangle$ of order 3^3. The index p_1 is 3, so we need to extend M to a Sylow 3-subgroup of G. The element h_1 is a_1, and M^{a_1} is $\langle a_2a_5^6, a_3a_6^5, a_4\rangle$. We call *conj_elt* to find the element y conjugating M^{a_1} to M.

In *conj_elt*, the recursion works up the chain of subnormal subgroups considering the values $j = 4, 3, 2$. For $j = 4$, it decides that the identity conjugates $\langle a_4\rangle$ to $\langle a_4\rangle$. For $j = 3$, it decides that a_6^5 conjugates $\langle a_3a_6^5, L\rangle$ to $\langle a_3, L\rangle$, where $L = \langle a_4\rangle$, because $h = a_3a_6^5$, $h^{-1}k_1 = a_6^2$ of order 7, $m = 0$, $q = 7$, $a = 1$, $b = 2$, $x = a_6^2$ and $z = a_6^5$. For $j = 2$, it decides that a_5^6

conjugates $<a_2a_5^6, L>$ to $<a_2, L>$, where $L = <a_3,a_4>$, because $h = a_2a_5^6$, $h^{-1}k_1 = a_5$ of order 7, $m = 0$, $q = 7$, $a = 1$, $b = 2$, $x = a_5$ and $z = a_5^6$. Hence, the element y conjugating M^{a_1} to M is $a_5^6a_6^5$.

So *pc_sylow* returns

$$< a_1a_5^6a_6^5, a_2, a_3, a_4 >$$

as the Sylow 3-subgroup of G. As a permutation,

$$a_1a_5^6a_6^5 = (1,11,15)(2,10,12)(3,14,19)(4,21,20)(5,13,18)(6,16,17)(7,8,9).$$

Summary

This chapter has defined the concepts associated with pc presentations for p-groups and soluble groups. It has briefly described a few algorithms based on pc presentations.

Exercises

(1/Easy) Determine all the conditioned pc presentations (as a finite soluble group) of the symmetries of the square.

(2/Easy) Determine all the conditioned pc presentations (as a finite p-group) of the symmetries of the square.

(3/Moderate) Consider the given pc presentation for S_4, the symmetric group of degree 4. Use a suitably modified form of the non-commutative Gaussian algorithm to compute a cgs of $<a_1, a_4>$, a cgs of $<a_2,a_1a_2^2a_4>$, and a cgs for $<a_2,a_3a_4>$.

(4/Easy) Compute the conjugacy classes of the symmetries of the square.

(5/Easy) Compute a Sylow 2-subgroup of S_4.

Bibliographical Remarks

The use of Cayley graphs and presentations to represent groups is widespread in combinatorial group theory. The correspondence between loops and relations is also well known. The classic books H.S.M. Coxeter and W.O.J. Moser, **Generators and Relations for Discrete Groups**, Springer-Verlag, Berlin, 1965, and W. Magnus, A. Karass, and D. Solitar, **Combinatorial Group Theory**, Interscience, New York, 1966 contain much more on Cayley graphs, relations, and presentations.

A vast range of algorithms for p-groups and soluble groups have been developed, implemented, and described by R. Laue, J. Neubüser, and U. Schoenwaelder, *"Algorithms for finite soluble groups and the SOGOS system"*, **Computational Group Theory**, M.D. Atkinson, (ed.), Academic Press, London, 1984, pp.105-135. They either use the approach of the conjugacy class algorithm of extending a result from $G/G(i)$ to $G/G(i+1)$, or consider the action of subgroups and compute appropriate orbits and stabilisers (as in the normalizer algorithm). This is a good paper to read as an introduction to the area.

Further work on algorithms for soluble groups and p-groups has been done by J.J. Cannon and C.R. Leedham-Green in forthcoming publications, and by Glasby, Slattery, Conlon, and Neubüser and Mecky in S.P. Glasby, *"Constructing normalisers in finite soluble groups"*, J. Symbolic Computation **5** (1988) 285-294; S.P. Glasby and M.C. Slattery, *"Computing* *intersections and normalizers in soluble groups"*, J. Symbolic Computation **9** (1990) 637-651; S.B. Conlon, *"Computing modular and projective character degrees of soluble groups"*, J. Symbolic Computation **9** (1990) 551-570; S.B. Conlon, *"Calculating characters of p-groups"*, J. Symbolic Computation **9** (1990) 535-550; M. Mecky and J. Neubüser, *"Some remarks on the computation of conjugacy classes of soluble groups"*, Bulletin Australian Math. Soc. **40** (1989) 281-292.

AG-systems for 2-groups were implemented by J. Neubüser, *"Bestimmung der Untergruppenverbände endlicher p-Gruppen auf einer programmgesteuerten elektronischen Dualmaschine"*, Numerische Mathematik **3** (1961) 271-278; and for general finite soluble groups by Lindenberg and Jürgensen in W. Lindenberg, *"Über eine Darstellung von Gruppenelementen in digitalen Rechenautomaten"*, Numerische Mathematik **4** (1962) 151-153; W. Lindenberg, *"Die Struktur eines Übersetzungsprogramm zur Multiplikation von Gruppenelementen in digitalen Rechenautomaten"*, Mitteilungen des Rhein-Westfälischen Institut für Instrumentalische Mathematik Bonn **2** (1963) 1-38; and H. Jürgensen, *"Calculation with the elements of a finite group given by generators and defining relations"*, **Computational Problems in Abstract Algebra**, J. Leech (ed.), Pergamon Press, Oxford, 1970, pp.47-57. There is no universal agreement on notation and terminology. We follow the terminology for pc presentations in M.F. Newman, *"Calculating presentations for certain kinds of quotient groups"*, **SYMSAC '76** (Proc. 1976 ACM Symp. on Symbolic and Algebraic Computation, Yorktown Heights, 1976), R.D. Jenks (ed.), ACM, New York, 1976, pp.2-8.

An introduction to rewriting is B. Buchberger and R. Loos, *"Algebraic simplification"*, **Computer Algebra: Symbolic and Algebraic Computation**, (2nd edition), B. Buchberger, G.E. Collins, R.G.K. Loos (editors), Springer Verlag, Wien, 1983, pp.11-43.

Collection has a long history, going back to theoretical work of P. Hall, *"A contribution to the theory of groups of prime-power order"*, Proceedings of the London Mathematical Society (2) **36** (1934) 29-95. The first implementations followed Hall and used collection to the left: H. Felsch, **Die Behandlung zweier gruppen-theoretischer Verfahren auf elektronischen Rechenmaschinen**, Diplomarbeit, Kiel, 1960; J.M. Campbell and W.J. Lamberth, *"Symbolic and numeric computation in group theory"*, Proceedings of the Third Australian Computer Conference (Canberra, 1966), Australian Trade Publications, Chippendale, Australia, 1967, pp.293-296; E. Czyzo, *"An attempt of mechanization of Hall collecting process"*, Algorytmy **9** (1972) 5-17; E. Czyzo, *"An automatization of the commutator calculus"*, Algorytmy **10** (1973) 23-34; I.D. Macdonald, *"A computer application to finite p-groups"*, Journal of the Australian Mathematical Society **17** (1974) 102-112; and W. Felsch, **Ein commutator collecting Algorithmus zur Bestimmung einer Zentralreihe einer endlichen p-Gruppe**, Staatsexamensarbeit, Aachen, 1974. In 1961, Neubüser discovered that collection from the right was superior in terms of space and time usage. This strategy was used in the implementations of Neubüser, Lindenberg, Jürgensen cited above, and by T.W. Sag and J.W. Wamsley, *"On computing the minimal number of defining relations for finite groups"*, Mathematics of Computation **27** (1973) 361-368; J.W. Wamsley, *"Computation in nilpotent groups (theory)"*, (Proceedings of the 2nd International Conference on the Theory of Finite Groups, Canberra, 1973), M.F. Newman (editor), Lecture Notes in Mathematics **372**, Springer-Verlag, Berlin, 1974, pp.691-700; V. Felsch, *"A machine independent implementation of a collection algorithm for multiplication of group elements"*, **SYMSAC '76** (Proc. 1976 ACM Symp. on Symbolic and Algebraic Computation, Yorktown Heights,

1976), R.D. Jenks (ed.), ACM, New York, 1976, pp.159-166; and G. Havas and T. Nicholson, *"Collection"*, **SYMSAC '76** (Proc. 1976 ACM Symp. on Symbolic and Algebraic Computation, Yorktown Heights, 1976), R.D. Jenks (ed.), ACM, New York, 1976, pp.9-14. Collection from the right was independently discovered by Wamsley in 1972 in association with the nilpotent quotient algorithm (NQA). An important variation to this strategy is *combinatorial collection* for *p*-groups introduced by G. Havas and T. Nicholson, *"Collection"*, **SYMSAC '76** (Proc. 1976 ACM Symp. on Symbolic and Algebraic Computation, Yorktown Heights, 1976), R.D. Jenks (ed.), ACM, New York, 1976, pp.9-14. Another useful idea is that of pre-compiling or pre-processing the presentation if many collections are to be performed. A detailed description of this approach is V. Felsch, *"A machine independent implementation of a collection algorithm for multiplication of group elements"*, **SYMSAC '76** (Proc. 1976 ACM Symp. on Symbolic and Algebraic Computation, Yorktown Heights, 1976), R.D. Jenks (ed.), ACM, New York, 1976, pp.159-166. Collection from the left is championed in the work of C.R. Leedham-Green and L.H. Soicher, *"Collection from the left and other strategies"*, J. Symbolic Computation **9** (1990) 665-675. They present empirical and theoretical evidence that it is the best strategy for general use. In connection with the NQA, their conclusion is supported by the empirical evidence of M.R. Vaughan-Lee, *"Collection from the left"*, J. Symbolic Computation **9** (1990) 725-733, which also contains a detailed description of an implementation of collection. The termination of the collection process and the concept of a collected ordering is discussed in C.C. Sims, *"Verifying nilpotence"*, J. Symbolic Computation **3** (1987) 231-247.

Echelonization and canonical generating sequences of subgroups are due to M.F. Newman in unpublished work in 1977. The first implementation and description appeared in R. Laue, J. Neubüser, and U. Schoenwaelder, *"Algorithms for finite soluble groups and the SOGOS system"*, **Computational Group Theory**, M.D. Atkinson, (ed.), Academic Press, London, 1984, pp.105-135.

The first application of canonical generating sequences in a top-down approach was the determination of the conjugacy classes of elements of a *p*-group by V. Felsch and J. Neubüser, *"An algorithm for the computation of conjugacy classes and centralizers in p-groups"*, **EUROSAM '79** (Proc. European Symp. on Symbolic and Algebraic Manipulation, Marseille, June 26-28, 1979), E.W. Ng (ed.), Lecture Notes in Computer Science **72**, Springer-Verlag, Berlin, 1979, 452-465. It was noted in the SOGOS paper of 1984 that the algorithm could be extended to finite soluble groups described by a pc presentation. This was later done as part of SOGOS in 1986 by M. Mecky and later refined by M. Mecky and J. Neubüser, *"Some remarks on the computation of conjugacy classes of soluble groups"*, Bulletin Australian Math. Soc. **40** (1989) 281-292. An algorithm for the classes of a soluble group was also implemented by J.J. Cannon and M.C. Slattery in 1987 in the Cayley system.

The work on conjugacy classes of elements can be generalized to determine the characters of the group (or at least the degrees of the characters). See S.B. Conlon, *"Computing modular and projective character degrees of soluble groups"*, J. Symbolic Computation **9** (1990) 551-570; S.B. Conlon, *"Calculating characters of p-groups"*, J. Symbolic Computation **9** (1990) 535-550.

The algorithm for computing Sylow subgroups of a soluble group and its proof of correctness is presented in S.P. Glasby, *"Constructing normalisers in finite soluble groups"*, J. Symbolic Computation **5** (1988) 285-294. The algorithm follows ideas of W.M. Kantor.

For a p-group given by an arbitrary presentation, the nilpotent quotient algorithm (NQA) will produce a pc presentation. In fact it does more than this: Given an arbitrary presentation of a group G, a prime p, and an integer bound c on the class (that is, the length of the lower exponent-p central series), it returns a pc presentation for the largest quotient P of G which is a p-group and has class at most c. The NQA is due to I.D. Macdonald, who in 1971 had an implementation of the NQA using collection to the left. His version of the algorithm worked down the lower central series and extended the quotient by elementary abelian steps. See I.D. Macdonald, *"A computer application to finite p-groups"*, Journal of the Australian Mathematical Society **17** (1974) 102-112. J.W. Wamsley noticed the benefits of collection from the right, and together with Bayes and Kautsky implemented the algorithm in 1973 - see J.W. Wamsley, *"Computation in nilpotent groups (theory)"*, (Proceedings of the 2nd International Conference on the Theory of Finite Groups, Canberra, 1973), M.F. Newman (editor), Lecture Notes in Mathematics **372**, Springer-Verlag, Berlin, 1974, pp.691-700; A.J. Bayes, J. Kautsky, J.W. Wamsley, *"Computation in nilpotent groups (application)"*, (Proceedings of the 2nd International Conference on the Theory of Finite Groups, Canberra, 1973), M.F. Newman (editor), Lecture Notes in Mathematics **372**, Springer-Verlag, Berlin, 1974, pp.82-89. The best implementation to date is due to W.A. Alford, G. Havas, and M.F. Newman. Havas and Newman introduced many ideas such as using the lower exponent-p-central series, and extending the quotient by steps of size p. This work is described in M.F. Newman, *"Calculating presentations for certain kinds of quotient groups"*, **SYMSAC '76** (Proc. 1976 ACM Symp. on Symbolic and Algebraic Computation, Yorktown Heights, 1976), R.D. Jenks (ed.), ACM, New York, 1976, pp.2-8, and G. Havas and M.F. Newman, *"Application of computers to questions like those of Burnside"*, **Burnside Groups**, Lecture Notes in Mathematics **806**, Springer-Verlag, Berlin, 1980, pp.211-230. The implementation incorporates contributions of M.R. Vaughan-Lee, *"An aspect of the nilpotent quotient algorithm"*, **Computational Group Theory**, M.D. Atkinson, (ed.), Academic Press, London, 1984, pp.75-83, on reducing the number of critical pairs that must be checked in order to ensure consistency - that is, list (C').

Some of the algorithms based on echelonization of subgroup generators are analogues of algorithms from linear algebra. One such is the (original) algorithm to compute intersections, which is an analogue of the algorithm for finding the intersection of vector subspaces (see H.G. Zimmer, **Computational Problems, Methods, and Results in Algebraic Number Theory**, Lecture Notes in Mathematics **268**, Springer-Verlag, Berlin, 1972, pp. 24).

Chapter 18. Soluble Permutation Groups

This chapter deals with both soluble permutation groups and the special case of permutation p-groups. Both the general and special case rely on a data structure which represents both the subnormal series associated with a pc presentation and the base and strong generating sets of the subgroups in the series. The main concern is constructing the data structure from the permutation group representation - and the isomorphism between the permutation group and the group described by the pc presentation.

Combining Series Generators and Strong Generators

Let G be a finite soluble permutation group and let $B = [\beta_1, \beta_2, ..., \beta_k]$ be a base for G. A sequence $\vec{g} = [g_1, g_2, ..., g_m]$ of elements of G is called a *B-strong series-generating sequence* (B-ssgs) if

1. $\vec{g}$ is a series-generating sequence (gs) for a subnormal series
$$\{identity\} = G(m+1) \lhd G(m) \lhd \cdots \lhd G(2) \lhd G(1) = G,$$
of G, where $G(i) = \langle g_i, g_{i+1}, ..., g_m \rangle$; and
2. each set $\{g_i, g_{i+1}, ..., g_m\}$ is a strong generating set of $G(i)$ relative to B.

For example, the symmetries of the square has a base B=[1,3] and a B-ssgs given by g_1=(3,4), g_2=(1,2)(3,4), and g_3=(1,3)(2,4). The symmetric group of degree 4 has a base B=[1,2,3] and a B-ssgs given by g_1=(3,4), g_2=(2,3,4), g_3=(1,3)(2,4), and g_4=(1,2)(3,4).

Given a base B, we can always find a B-ssgs. Since the group G is soluble, we can find a prime-step subnormal series
$$\{identity\} = G(m+1) \lhd G(m) \lhd \cdots \lhd G(2) \lhd G(1) = G,$$
and choose g_i to be the first permutation in $G(i) - G(i+1)$, under the lexicographical ordering on G induced from the ordering of Ω where the base points $\beta_1, \beta_2, ..., \beta_k$ are the first k points. These elements form a B-ssgs $[g_i : i = 1,2,...,m]$.

Generally, we work with a prime-step subnormal series where the index $|G(i):G(i+1)|$ is a prime p_i. A general B-ssgs is easily refined to a prime-step one by inserting appropriate powers of a generator g_i whenever the index $|G(i):G(i+1)|$ is composite.

Suppose we are working with a subgroup H of a soluble permutation group G. Suppose B is a base for G, and we know a B-ssgs $\vec{g}$ of H. A common method of constructing subgroups K is the cyclic extension method where $K = \langle H,y \rangle$ and the element $y \notin H$ normalizes H and satisfies $y^p \in H$, for some prime p. Of course, we wish to construct a B-ssgs of K. The subnormal series of K is
$$\{identity\} = H(m+1) \lhd H(m) \lhd \cdots \lhd H(2) \lhd H(1) = H \lhd K,$$
B is a base for K, and $\vec{g}$ contains the necessary strong generators of $H(i)$, so the outstanding

problem is to choose $g_0 \in K$ such that the set $\{g_0, g_1, ..., g_m\}$ is a strong generating set of K relative to B. To do this, we choose $g_0 \in yH$ as low in the stabiliser chain as possible. The stripping process of testing $y \in H$ will return a residue g_0 with this property. This is summarised in Algorithm 1.

Algorithm 1 : Prime-step Normalizing Generator

Input: a permutation group H with base $B=[\beta_1, \beta_2, \ldots, \beta_k]$,
strong generating set T and Schreier vectors $v^{(i)}$;
a prime p;
a permutation $y \notin H$ normalizing H such that $y^p \in H$;

Output: a base for $K = \langle H, y \rangle$ which may extend B;
an element g_0 such that $T \cup \{g_0\}$ is a strong generating set of K;

```
begin
  g_0 := y;  extend_base := true;  i := 0;
  while i < k and extend_base do
    i := i + 1;
    if β_i^{g_0} ∈ Δ^(i) then
      while β_i^{g_0} ≠ β_i do
        g_0 := g_0 × v^(i) [β_i^{g_0}]^{-1};
      end while;
    else
      extend_base := false;
    end if;
  end while;

  if extend_base then
    choose a point β_{k+1} moved by g_0;
    append β_{k+1} to B;
  end if;
end.
```

As an example of Algorithm 1, consider the symmetric group of degree 4, where $H=\langle g_2, g_3, g_4 \rangle$ is the alternating group and $y=(1,2)$; Then $K=\langle H, y \rangle$ is S_4 and the algorithm returns $g_0=(3,4)$.

The algorithm can be generalized to handle the case where y normalizes H and we do not know that the index $|K:H|$ is prime. The generalized algorithm will find a sequence of elements extending the strong generating set of H to a strong generating set of K. The details are in Algorithm 2.

Algorithm 2 : Normalizing Generator

Input: a permutation group H with base $B=[\beta_1,\beta_2,\ldots,\beta_k]$,
strong generating set T, Schreier vectors $v^{(i)}$, and basic orbits $\Delta^{(i)}$;
a permutation y normalizing H;

Output: a base for $K = \langle H,y\rangle$ which may extend B;
a sequence of elements $[g_1,g_2,\ldots,g_r]$ such that, for each j,
(a) $T \cup \{g_j,g_{j+1},\ldots,g_r\}$ is a strong generating set of $\langle H,g_j,g_{j+1},\ldots,g_r\rangle$,
(b) $g_j \notin \langle H,g_{j+1},g_{j+2},\ldots,g_r\rangle$,
(c) g_j normalizes $\langle H,g_{j+1},g_{j+2},\ldots,g_r\rangle$, and
(d) $g_j^{p_j} \in \langle H,g_{j+1},g_{j+2},\ldots,g_r\rangle$, for some prime p_j;

begin
  $z := y$; $r := 0$; $i := 0$;
  while $i < k$ **do**
    $i := i+1$;
    if $\beta_i^z \in \Delta^{(i)}$ **then**
      while $\beta_i^z \neq \beta_i$ **do** $z := z \times v^{(i)}[\beta_i^z]^{-1}$; **end while**;
    else
      let $p_1^{n_1} p_2^{n_2} \cdots p_t^{n_t}$ be the prime factorisation of $|\beta_i^{\langle H^{(i)},z\rangle}| \,/\, |\Delta^{(i)}|$;
      for $j := 1$ **to** t **do**
        for $l := 1$ **to** n_j **do**
          $r := r+1$; $g_r := z$; $z := z^{p_j}$;
        end for;
      end for;
    end if;
  end while;
  while $z \neq$ *identity* **do**
    $k := k+1$;
    choose a point β_k moved by z;
    append β_k to B;
    let $p_1^{n_1} p_2^{n_2} \cdots p_t^{n_t}$ be the prime factorisation of the orbit length $|\beta_k^{\langle z\rangle}|$;
    for $j := 1$ **to** t **do**
      for $l := 1$ **to** n_j **do**
        $r := r+1$; $g_r := z$; $z := z^{p_j}$;
      end for;
    end for;
  end while;
end.

As an example of Algorithm 2, let H to be trivial subgroup, let $y=(1,2)(3,4,5,6)$ of order 4 and degree 6, and form $K = \langle y\rangle$. The first **while**-loop, which performs the stripping, does nothing in this example. The second **while**-loop could choose $\beta_1=1$, for which $|\beta_1^{\langle y\rangle}| = 2$, and $g_1=y=(1,2)(3,4,5,6)$. It would then perform a second iteration with $z=y^2=(3,5)(4,6)$, choose $\beta_2=2$ and $g_2=y^2=(3,5)(4,6)$. So, a base for K is $B=[1,2]$, and a B-ssgs is $[g_1,g_2]$.

Cyclically Extended Schreier Vectors

The advantage of a subnormal series is that elements of the group can be expressed as normal words. If we have a pcp then the collection process computes normal words, but if we are in a permutation representation how is the normal word computed? A base, strong generating set, and Schreier vectors allow us to compute a word in the strong generators for a permutation.

Lemma

Suppose $\vec{g}$ is a B-ssgs for a soluble permutation group G. Let $g \in G$ and suppose that w_1 and w_2 are words in the strong generators $\{g_1, g_2, ..., g_m\}$ for g. Then (#occurrences of g_1 in w_1) = (#occurrences of g_1 in w_2) mod p_1.

In particular, we can calculate the exponent ε_1 of g_1 in the normal word for g from any word for g. Iterating with $g_1^{-\varepsilon_1} \times g$ will determine ε_2, and so on.

As an application of the Lemma, consider the element g=(1,2,3,4) in the symmetric group of degree 4. As $g = g_1 \times g_4 \times g_1 \times g_3 \times g_1 \times g_1 \times g_2 \times g_4 \times g_4 \times g_1$, the exponent $\varepsilon_1 = 1$. Then $g_1^{-\varepsilon_1} \times g$ = (1,2,3) satisfies the word $g_3 \times g_4 \times g_2 \times g_2 \times g_3 \times g_4$, so the exponent $\varepsilon_2 = 2$. Then $g_2^{-\varepsilon_2} \times g_1^{-\varepsilon_1} \times g$ = (1,2)(3,4) = g_4, so $\varepsilon_3 = 0$, and $\varepsilon_4 = 1$. Hence, the normal word for g is $g_1^1 \times g_2^2 \times g_3^0 \times g_4^1$.

The next data structure to be described combines the Schreier vectors of all the subgroups in the subnormal series into just one set of Schreier vectors. Since g_i normalizes $G(i+1)$ and $g_i^{p_i} \in G(i+1)$ then

$$\{identity, g_i, g_i^2, ..., g_i^{p_i - 1}\}$$

is a set of coset representatives for $G(i+1)$ in $G(i)$. Furthermore, suppose that l=$level(g_i)$ is the level in the stabilizer chain which contains g_i. That is, $g_i \in G^{(l)} - G^{(l+1)}$. Define a set $U(i,l)$ of coset representatives of $G(i)^{(l+1)}$ in $G(i)^{(l)}$ recursively by

$$\{ \ u \times g_i^j \ \mid \ u \in U(i+1,l), 0 \le j < p_i \ \},$$

and

$$U(m+1,l) = \{ \ identity \ \}, \text{ for all } l.$$

We say that g_i *cyclically extends* $U(i+1,l)$ to $U(i,l)$.

A *cyclically extended Schreier vector* (cesv) is a Schreier vector $v^{(l)}$ that represents the set $U(i,l)$ for each i, $1 \le i \le m+1$.

Consider again the example of S_4 with base B=[1,2,3] and B-ssgs $[g_1,g_2,g_3,g_4]$. The cyclically extended Schreier vectors are

	1	2	3	4
$v^{(1)}$		4	3	3
$v^{(2)}$			2	2
$v^{(3)}$				1

Note that we can think of the Schreier vector entries as either being the generator g_i, or just the index i of the generator in the B-ssgs. We have chosen the latter in this case, but it is often convenient to allow either notation.

By restricting to those entries which refer only to the generators $\{g_i, g_{i+1}, \ldots, g_m\}$, we have a set of Schreier vectors for the group $G(i)$. The diagrams below show the set of Schreier vectors for each group in the subnormal series of S_4 given by the B-ssgs $[g_1, g_2, g_3, g_4]$.

Cyclically Extended Schreier Vectors of S_4

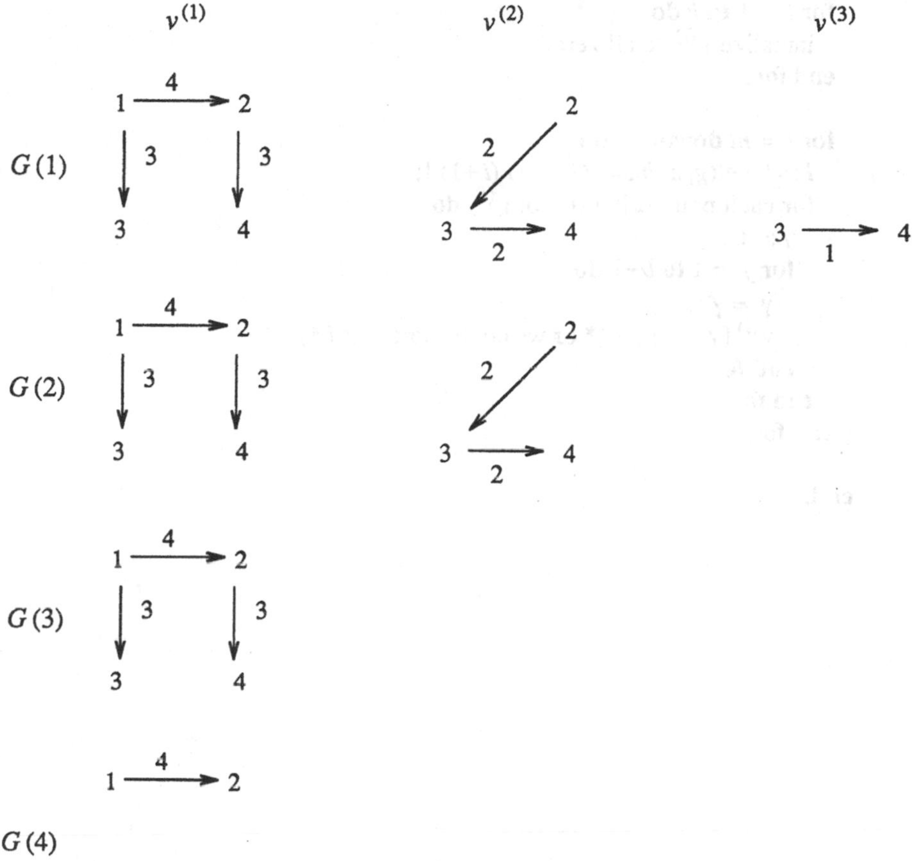

The cyclically extended Schreier vectors can be calculated directly from the definition, by working up the subnormal series, as done in Algorithm 3. The cesv's can be used, as in Algorithm 4, to compute a normal word of an element g of the group G.

Algorithm 3 : Construct Cyclically Extended Schreier Vectors

Input : a base $B = [\beta_1, \beta_2, ..., \beta_k]$ for a group G;
a B-ssgs $[g_1, g_2, ..., g_m]$ for G;

Output : cyclically extended Schreier vectors $v^{(i)}$, i=1,2,...,k of G;

begin

for l := 1 **to** k **do**
initialize $v^{(l)}$ to all zero;
end for;

for i := m **downto** 1 **do**
l := level(g_i); $b := |G(i) : G(i+1)|$;
for each point α in orbit of $v^{(l)}$ **do**
$\gamma := \alpha$;
for j := 1 **to** $b-1$ **do**
$\gamma := \gamma^{g_i}$;
$v^{(l)}[\gamma] := g_i$; (* or we could store just i *)
end for;
end for;
end for;

end.

Algorithm 4 : Normal Word

Input : a soluble permutation group G;
a base $B = [\beta_1, \beta_2, ..., \beta_k]$ for a group G;
a B-ssgs $[g_1, g_2, ..., g_m]$ for G;
cyclically extended Schreier vectors $v^{(i)}$ of G;
an element g of G;

Output : the exponents $[\varepsilon_1, \varepsilon_2, \cdots, \varepsilon_m]$ of the normal word of g;

```
begin

  for i := 1 to m do
    l := level(g_i);
    (* strip g to level l *)
    h := g;
    for j := 1 to l-1 do
      α := β_j^h;
      while α ≠ β_j do
        h := h × v^(j) [α]^-1;(* entry of cesv as a generator *)
        α := β_j^h;
      end while ;
    end for;
    (* determine exponent ε_i *)
    ε_i := 0;  γ := β_l^h;
    while v^(l) [γ] = i do (* entry of cesv as an index *)
      ε_i := ε_i+1; γ := γ^(g_i^-1);
    end while;
    (* residue is now in G (i+1) *)
    g := g_i^(-ε_i) × g;
  end for;

end.
```

There are occasions where we require a subnormal series to satisfy additional properties. The subnormal series will be obtained by modifying an existing B-ssgs in an iterative way. The simplest modification is where a B-ssgs $\vec{g} = [g_1, g_2, ..., g_m]$ is changed to $\bar{g} = [g_1, g_2, ..., g_{l-1}, h, g, g_{l+2}, g_{l+3}, ..., g_m]$. Of course, the elements g and h are chosen so that $\bar{g}$ is a B-ssgs. In the simplest case we may also choose them to lie in $G(l)$. Let $\bar{G}(i)$ denote the i-th term of the subnormal series defined by $\bar{g}$. Then $G(i) = \bar{G}(i)$, for all i except $i = l+1$. Hence, when we modify the cesv's to correspond to the new B-ssgs, there is very little to do. The details are in Algorithm 5.

Algorithm 5 : Changing a B-ssgs

Input : a soluble permutation group G with a base $B = [\beta_1, \beta_2, ..., \beta_k]$;
  a B-ssgs $\vec{g} = [g_1, g_2, ..., g_m]$ and cesv's $v^{(i)}$ of G;
  an integer l, and elements g,h in $G(l)$ such that
    $G(l) = \langle g, h, g_{l+2}, g_{l+3}, ..., g_m \rangle$
    and $[g_1, g_2, ..., g_{l-1}, h, g, g_{l+2}, g_{l+3}, ..., g_m]$ is a B-ssgs of G;
Output : cyclically extended Schreier vectors relative to new B-ssgs;
begin
  i := level(g_l); i_1 := level(g_{l+1});
  if $i = i_1$ **then**
    $b := |\bar{G}(l+1):G(l+2)|$; (* for new group $\bar{G}(l+1)$ *)
    (* Run through Schreier vector noting effect of g, the new g_{l+1} *)
    for each α in Ω **do**
      if $v^{(i)}[\alpha] \geq l+2$ **then**
        for j := 1 **to** $b-1$ **do** $v^{(i)}[\alpha^{g^j}]$:= '*'; **end for;**
      end if;
    end for;
    (* Run through noting effect of h, the new g_l, and correcting entries *)
    for each α in Ω **do**
      if $v^{(i)}[\alpha] = l+1$ *or* l **then** $v^{(i)}[\alpha] := l$; **end if;**
      if $v^{(i)}[\alpha]$ = '*' **then** $v^{(i)}[\alpha] := l+1$; **end if;**
    end for;
  else (* One of g or h belongs to level i, and the other belongs to level i_1 *)
    if level(h) = i_1 **then** (* orbits don't change, just labels *)
      for each α in Ω **do**
        if $v^{(i_1)}[\alpha] = l+1$ **then** $v^{(i_1)}[\alpha] := l$; **end if;**
        if $v^{(i)}[\alpha] = l$ **then** $v^{(i)}[\alpha] := l+1$; **end if;**
      end for;
    end if;
  end if;
end.

For example, the symmetries of the square has a base B=[1,3] and a B-ssgs $\vec{g}$ given by g_1=(3,4), g_2=(1,2)(3,4), and g_3=(1,3)(2,4). The cyclically extended Schreier vectors are

	1	2	3	4
$v^{(1)}$		2	3	2
$v^{(2)}$				1

Changing the B-ssgs to $\bar{g}$ = [(3,4), (1,3)(2,4), (1,4)(2,3)] gives the cesv's

	1	2	3	4
$v^{(1)}$		2	2	3
$v^{(2)}$				1

Changing the original B-ssgs $\vec{g}$ to $\overline{g}$ = [(3,4), (1,3)(2,4), (1,2)(3,4)] gives the cesv's

	1	2	3	4
$v^{(1)}$		3	2	2
$v^{(2)}$				1

Changing this B-ssgs to $\overline{g}$ = [(1,3)(2,4), (3,4), (1,2)(3,4)] gives the cesv's

	1	2	3	4
$v^{(1)}$		3	1	1
$v^{(2)}$				2

Using Homomorphisms

This section considers the interaction between homomorphisms and a gs or B-ssgs of a soluble group G. The first result follows directly from the fact that the kernel of a homomorphism is a normal subgroup.

Lemma

Let $f : G \longrightarrow H$ be a homomorphism. If $[g_1, g_2, \ldots, g_t]$ is a series-generating sequence for $ker(f)$, and $[f(h_1), f(h_2), \ldots, f(h_r)]$ is a series-generating sequence for $im(f)$, then $[h_1, h_2, \ldots, h_r, g_1, g_2, \ldots, g_t]$ is a series-generating sequence for G.

If both gs's define a prime-step series then so does the resulting gs $[h_1, h_2, \ldots, h_r, g_1, g_2, \ldots, g_t]$. A B-ssgs can be constructed from this sequence by repeated calls to Algorithm 1. This construction is presented in Algorithm 6.

Algorithm 6 : B-ssgs using Homomorphism

Input: soluble permutation groups G and H;
a base B for G;
a homomorphism $f : G \longrightarrow H$;
a prime-step gs $[g_1, g_2, \ldots, g_t]$ of $ker(f)$;
a prime-step gs $[f(h_1), f(h_2), \ldots, f(h_r)]$ of $im(f)$;
Output: a B-ssgs of G;
begin
$K := \langle identity \rangle$;
for $i := t$ **downto** 1 **do**
$K := \langle K, g_i \rangle$ using Algorithm 1;
end for;
(* now have B-ssgs of ker(f) *)

for $i := r$ **downto** 1 **do**
$K := \langle K, h_i \rangle$ using Algorithm 1;
end for;
(* now have B-ssgs of G *)
end.

The transitive constituent homomorphism $f{:}G \longrightarrow G\,|_\Delta$, where $\Delta \subseteq \Omega$ is invariant under the soluble permutation group G, interacts nicely with B-ssgs's of the image and kernel. The following theorem gives the details.

Theorem

Let G be a soluble permutation group. Let Δ be a proper subset of Ω which is invariant under G. Let $f : G \longrightarrow G\,|_\Delta$ be the natural constituent homomorphism, and let $\bar{f} : \Delta \longrightarrow \{1,2,...,|\Delta|\}$ be the associated relabelling of points of Δ. Suppose $A = [\bar{f}(\alpha_1), \bar{f}(\alpha_2), ..., \bar{f}(\alpha_k)]$ is a base for $im(f)$ and $[f(h_1), f(h_2), ..., f(h_r)]$ is a prime-step A-strong series-generating sequence of $im(f)$. Further suppose that $B = [\beta_1, \beta_2, \ldots, \beta_l]$ is a base for $ker(f)$ chosen from $\Omega - \Delta$ and that $[g_1, g_2, \ldots, g_t]$ is a prime-step B-strong series-generating sequence of $ker(f)$. Let $AB = [\alpha_1, \alpha_2, \cdots, \alpha_k, \beta_1, \cdots, \beta_l]$. Then AB is a base for G and $[h_1, h_2, ..., h_r, g_1, g_2, ..., g_t]$ is a prime-step AB-strong series-generating sequence of G.

The blocks homomorphism has similar nice properties when G is a p-group and the block system is maximal. In this situation the image is a cyclic group of order p permuting the p blocks of the system as a p-cycle. Hence, the preimage of this p-cycle extends a B-ssgs of $ker(f)$ to a strong generating set of G (and hence to a B-ssgs of G).

Theorem

Let G be a transitive permutation p-group. Let π be a system of imprimitivity of G with p blocks. Let $f : G \longrightarrow G\,|_\pi$ be the natural blocks homomorphism. Let $f(g)$ be a generator of $im(f)$. Suppose $B = [\beta_1, \beta_2, \ldots, \beta_k]$ is a base for $ker(f)$ and that $[g_1, g_2, \ldots, g_r]$ is a B-strong series-generating sequence of $ker(f)$. Then B is a base for G, and $[g, g_1, g_2, ..., g_r]$ is a B-strong series-generating sequence of G.

The only primitive permutation p-group is the cyclic group of order p acting regularly on p points.

The result about blocks homomorphisms does not extend nicely to soluble groups. We can still apply the Lemma to construct a gs, and then use Algorithm 6 to establish the strong generating property.

Computing with the Isomorphism

A B-ssgs $\vec{g} = [g_1, g_2, ..., g_m]$ defines an isomorphism f between the soluble permutation group G and a group H defined by the pc presentation

$$< a_1, a_2, ..., a_m \mid a_i^{p_i} = \text{normal word of } g_i^{p_i},\ 1 \le i \le m,$$
$$a_j a_i = \text{normal word of } g_j g_i,\ 1 \le i < j \le m >,$$

where the normal words are written in terms of the generator symbols a_i rather than the B-ssgs generators g_i. The isomorphism is induced from its action on the generators, viz

$$f : G \longrightarrow H$$
$$g_i \longmapsto a_i$$

The computational tasks with the isomorphism are performed as follows:

1. Computing the image of the group: The image is H.

2. Computing the kernel: The kernel is the trivial subgroup of G.

3. Constructing the image of an element: Let $g \in G$. Form the normal word w of g using the B-ssgs and the cyclically extended Schreier vectors. Rewrite the word w in terms of the a_i to give the image $f(g) \in H$.

4. Constructing the preimage of an element: Let $h \in H$ be given by a normal word w in the a_i. Evaluate this word w using the permutations g_i to form $f^{-1}(h)$.

5. Constructing the image of a subgroup: Let $L \leq G$ be a subgroup generated by the set T of elements. Form $f(T) = \{ f(t) \mid t \in T \}$ and use non-commutative Gaussian elimination to form a canonical generating sequence (cgs) of the subgroup $f(L) = \langle f(T) \rangle$ of H.

6. Constructing the preimage of a subgroup: Let $L \leq H$ be given by a cgs $[c_1, c_2, ..., c_r]$. A B-ssgs for $K = f^{-1}(L)$ is formed by

$K := \langle$ *identity* $\rangle$;

for $i := r$ **downto** 1 **do**
  $K := \langle K, f^{-1}(c_i) \rangle$ using Algorithm 1;
end for;

As an example of 6, consider the symmetric group of degree 4 with $B = [1,2,3]$ and $\vec{g} = [g_1, g_2, g_3, g_4]$. The subgroup $L = \langle a_1 a_3, a_2 a_4 \rangle$ is isomorphic to S_3 and has cgs $[a_1 a_3, a_2 a_4]$. The preimages of these elements are $g_1 g_3 = (1,2)$ and $g_2 g_4 = (1,3,2)$. A B-ssgs of $K = f^{-1}(L)$ is [(2,3), (1,3,2)] because Algorithm 1 strips $g_1 g_3 = (1,2)$ by $g_2 g_4$ to give (2,3) (= $g_1 g_2$).

Subnormal Series: p-Groups

Given a permutation p-group G, we can directly apply the results on homomorphisms to determine a base B and a B-ssgs of G describing a prime-step subnormal series of G.

Algorithm 7 : B-ssgs of a p-group

Input : a permutation p-group G with a base and strong generating set;
Output : a base B for G and a prime-step B-ssgs $\vec{g}$ for G;

procedure BSSGS(G : group; **var** B : base; **var** $\vec{g}$: sequence of elements);
begin
  if $G = \langle identity \rangle$ **then**
    B := empty; $\vec{g}$:= empty;
  else if G has order p **then**
    let g be a generator of G;
    $B[1]$:= a point moved by g; $\vec{g}[1] := g$;
  else if G is transitive **then**
    find a maximal system π of imprimitivity;
    let f be the blocks homomorphism $G \longrightarrow G\,|_{\pi}$;
    BSSGS ($ker(f), B, \vec{g}$);
    $\vec{g} := [f^{-1}$(generator of $im(f))]$ **cat** $\vec{g}$;
  else
    find a non-trivial orbit Δ of G;
    let f be the constituent homomorphism $G \longrightarrow G\,|_{\Delta}$;
    let $\bar{f}$ be the associated relabelling $\Delta \longrightarrow \{1,2,\ldots,\Delta\}$;
    BSSGS ($ker(f), B, \vec{g}$); BSSGS ($im(f), A, \vec{h}$);
    $B := \bar{f}^{-1}(A)$ **cat** B; $\vec{g} := f^{-1}(\vec{h})$ **cat** $\vec{g}$;
  end if;
end (* BSSGS *);

begin
  B := empty; $\vec{g}$:= empty; BSSGS($G, B, \vec{g}$);
end.

As an example consider the symmetries G of the square generated by a=(1,4,3,2) and b=(2,4). A maximal system of imprimitivity is $\pi = \{1,3|2,4\}$. The image of the homomorphism f_π is generated by $f_\pi(a)$, and the kernel K is $\langle a^2, b \rangle$. The kernel K has orbits $\Delta = \{1,3\}$ and $\{2,4\}$. The image of the homomorphism f_Δ is $\langle f_\Delta(a^2) \rangle$, while the kernel is $\langle b \rangle$. Hence, the group G has a base B=[1,2] and a B-ssgs $[a,a^2,b]$.

Subnormal Series: Soluble Groups

A primitive soluble permutation group G has several very specific properties:

i. the degree of G is a prime power p^s;

ii. G has a subgroup N such that

   a. N acts regularly on Ω;

   b. N has order p^s;

   c. N is elementary abelian;

d. N is normal in G;

e. the point stabiliser G_1 is isomorphic to the quotient G/N;

f. every element $g \in G$ can be uniquely expressed as $g = a \times g'$, where $a \in N$ and $g' \in G_1$;

The subgroup N is called an *earns* of G - for elementary abelian regular normal subgroup.

For example, the symmetric group S_4 of degree 4 has the Klein four-group $<(1,2)(3,4),(1,3)(2,4)>$ as its earns.

Given a nonredundant generating set $\{a_1, a_2, \ldots, a_s\}$ of N, we can easily construct a B-ssgs of N by taking $B=[\beta]$ for any point $\beta \in \Omega$, and taking the generators in any order whatsoever.

Given any element $a \in N$, we can construct N as

$$< a^g \mid g \in G >,$$

the normal closure of $<a>$ in G. However, it is not always easy to locate an initial element $a \in N$.

The subgroup N can also be characterised as $O_p(G)$, the intersection of all Sylow p-subgroups of G. So if S is a Sylow p-subgroup of G, then N is constructed by Algorithm 8. However, this approach can be expensive.

Algorithm 8 : EARNS as $O_p(G)$

```
Input: a primitive soluble permutation group G of degree p^s;
       a Sylow p-subgroup S of G;
Output: the elementary abelian regular normal subgroup (earns) N of G;
begin
  N := S;   notnormal := true;
  while notnormal do
    notnormal := false;
    for each generator g of G do
      if N^g ≠ N then
        notnormal := true; N := N ∩ N^g; break;
      end if;
    end for;
  end while;
end.
```

Having constructed N and a base $B = [1]$ for N and a B-ssgs $\vec{a}$ of N, we can recursively compute a base A and an A-ssgs $\vec{g}$ of the point stabiliser G_1. The concatenation of these is a base $BA = B$ *cat* A and a BA-ssgs $\vec{g}$ *cat* $\vec{a}$ of G. The details are given in Algorithm 9. For example, let G be the symmetric group of degree 4. Then N is the Klein four-group $< g_3, g_4 >$. The point stabiliser has a base $A = [2,3]$ and A-ssgs $[g_1, g_2]$, giving a base $B = [1,2,3]$ for G and B-ssgs $[g_1, g_2, g_3, g_4]$.

Algorithm 9 : B-ssgs of a Soluble Group

Input : a soluble permutation group G with a base and strong generating set;

Output : a base B for G and a prime-step B-ssgs $\vec{g}$ for G;

```
procedure BSSGS( G : group; var B : base; var g⃗ : sequence of elements );

begin
 if G = <identity> then
     B := empty; g⃗ := empty;
 else if G is intransitive then
     find a non-trivial orbit Δ of G;
     let f be the constituent homomorphism G —→ G |Δ;
     let f̄ be the associated relabelling Δ —→ {1,2,...,Δ};
     BSSGS (ker(f), B, g⃗); BSSGS (im(f), A, h⃗);
     B := f̄⁻¹(A) cat B; g⃗ := f⁻¹(h⃗) cat g⃗;
 else if G is imprimitive then
     find a system π of imprimitivity;
     let f be the blocks homomorphism G —→ G|π;
     BSSGS (ker(f), B, g⃗); BSSGS (im(f), A, h⃗);
     K := ker(f);
     for i := length(h⃗) downto 1 do
        K := <K, h_i > using Algorithm 1;
     end for;
 else
     compute earns N and a [1]-ssgs a⃗ of N;
     BSSGS (G_1, B, h⃗);
     B := [1] cat B; g⃗ := h⃗ cat a⃗;
 end if;
end (* BSSGS *);

begin
  B := empty; g⃗ := empty; BSSGS( G, B, g⃗ );
end.
```

An alternative approach, which avoids costly constructions of the earns as $O_p(G)$, handles the costly cases by constructing a normal subgroup M of G of prime index in G. The computation of M uses homomorphisms and the structure of primitive soluble permutation groups. The computation is described in Algorithm 10, and its use in computing a B-ssgs is presented in Algorithm 11.

Algorithm 10 : Prime Index

```
Input  : a nontrivial soluble permutation group G
           with a base and strong generating set;

Output : a prime p;
         a normal subgroup M of G of index p;
         an element g ∈ G−M;

procedure prime_index( G : group ) : prime, subgroup, element;

begin
  if G is intransitive then
      find a non-trivial orbit Δ of G;
      let f be the constituent homomorphism G --> G |Δ;
      p, M, g := prime_index( f (G) );
      return p, f^-1(M), f^-1(g);
  else if G is imprimitive then
      find a maximal system π of imprimitivity;
      let f be the blocks homomorphism G --> G |π;
      p, M, g := prime_index( f (G) );
      return p, f^-1(M), f^-1(g);
  else
      H := G_1;
      if H = <identity> then (* G has prime order *)
        let g generate G;
        return |G|, <identity>, g;
      else if H has prime order then (* special case where N is easy to find *)
        let h generate H;
        find a generator g of G such that a := [g,h]≠ identity;
        N := < a, a^h, a^(h^2), ... >;
        return |H|, N, h;
      else
        p, L, h := prime_index( H );
        M := normal closure of L in G;
        return p, M, h;
      end if;
  end if;
end (* prime index *);
```

The normal closure computation in Algorithm 10 is particularly easy. The normal closure M is $N L = \{ n \times l \mid n \in N,\ l \in L \}$, where N is the earns, so the point stabiliser of M is L, and the order of M is $|L| \times |\Omega|$. Hence, it suffices to add conjugates x^g, where x is a generator of L and g is a generator of G to the strong generating set of L until the group is transitive. There is no need to perform a Schreier-Sims method as we already have a strong generating set of the point stabiliser.

Algorithm 11 : B-ssgs of a Soluble Group using Prime Index

Input : a soluble permutation group G with a base and strong generating set;
Output : a base B for G and a prime-step B-ssgs $\vec{g}$ for G;

```
procedure BSSGS( G : group; var B : base; var g⃗ : sequence of elements );
begin
 if G = <identity> then
     B := empty; g⃗ := empty;
 else if G is intransitive then
     find a non-trivial orbit Δ of G;
     let f be the constituent homomorphism G ⟶ G|Δ;
     let f̄ be the associated relabelling Δ ⟶ {1,2,...,Δ};
     BSSGS (ker(f), B, g⃗); BSSGS (im(f), A, h⃗);
     B := f̄⁻¹(A) cat B; g⃗ := f⁻¹(h⃗) cat g⃗;
 else if G is imprimitive then
     find a system π of imprimitivity;
     let f be the blocks homomorphism G ⟶ G|π;
     BSSGS (ker(f), B, g⃗); BSSGS (im(f), A, h⃗);
     K := ker(f);
     for i := length( h⃗ ) downto 1 do
       K := < K, h_i > using Algorithm 1;
     end for;
 else
     H := G_1;
     if H = <identity> then
       N := G;
       B := [1]; g⃗ := nonredundant generators of N in any order;
     else if H has prime order then (* special case where N is easy to find *)
       let h generate H; let β be a point moved by h;
       find a generator g of G such that a := [g,h] ≠ identity;
       N := < a, a^h, a^(h^2), ... >;
       a⃗ := nonredundant generators of N in any order;
       B := [ 1, β ]; g⃗ := [ h ] cat a⃗;
     else
       p, L, h := prime_index( H ); M := normal closure of L in G;
       BSSGS ( M, B, g⃗ );
       M := < M, h > using Algorithm 1;
     end if;
 end if;
end (* BSSGS *);

begin
  B := empty; g⃗ := empty; BSSGS( G, B, g⃗ );
end.
```

For example, let G be the symmetric group of degree 4. Algorithm 11 will construct $H = G_1$ isomorphic to S_3, $L = <g_2>$ of index 2 in H, and $h = (3,4)$. When calculating the normal closure, the conjugate $g_2^{g_3} = (1,2,4)$ is added to the strong generating set, giving $M = <(2,3,4),(1,2,4)>$ isomorphic to the alternating group A_4.

In the recursive call to *BSSGS* with A_4, the algorithm executes the special case. It finds the commutator $a = [(1,2,4),(2,3,4)] = (1,4)(2,3)$ and constructs the Klein four-group $<(1,4)(2,3),(1,2)(3,4)>$. Thus it returns a base B=[1,2] and a B-ssgs [(2,3,4), (1,4)(2,3), (1,2)(3,4)].

On return from the recursive call, the call to Algorithm 1 with h=(3,4) does not alter h, but it does extend the base B to [1,2,3]. The resulting B-ssgs is [(3,4), (2,3,4), (1,4)(2,3), (1,2)(3,4)].

Subnormal Series: Soluble Schreier Method

One can work directly from a generating set S of a soluble permutation group G to construct a B-ssgs. The algorithm comes straight from

a. the definition of a soluble group in that G has a proper normal subgroup N such that G/N is abelian (and N is soluble); and

b. the use of Algorithm 2 to construct a B-ssgs of the extensions of N formed during the execution of the algorithm. The conditions of Algorithm 2 are fulfilled because N is normal in G.

The algorithm constructs a subgroup M such that $N \leq M \leq G$. Eventually $M = G$. Given an element $y \in G-N$, the subgroup M is the normal closure of $<N,y>$ in G. During the course of this computation N may be updated, and at the conclusion of the computation M is normal in G.

The computation of M forms the normal closure (as usual) by considering the set U of conjugates of y under generators of G (and their iterated products). The properties of N are maintained so that $N \triangleleft M$ and M/N is abelian, by ensuring that all commutators $[u,v]$ with $u,v \in U$ are in N. If a commutator $w = [u,v]$ does not lie in N, then we recursively compute the normal closure of $<N,w>$ in G as the updated value of N. The details are presented as Algorithm 12.

Algorithm 12 : Soluble Schreier Method

```
Input: a soluble permutation group G given by generators S;
Output: a base B for G and a B-ssgs g⃗ of G;
begin
  N := < identity >;  B := empty;  g⃗ := empty;
  for each s in S do
    sol_normal_closure( G, N, s );
  end for;
  (* N is G *)
end.
```

```
procedure sol_normal_closure( G : group; var N : group; y : element of G );

(* Compute M, the normal closure of <N,y> in G where N ◁ G.
   We always have M/N is abelian during the course of the procedure.
   At the end of the procedure, M is returned as N. *)
begin
  M := N;  U := {};  Z := {y};
  while Z is not empty do
    choose z ∈ Z;  Z := Z - {z};
    if z ∉ M then
      T := U;  done := false;
      while T is not empty do
        choose u ∈ T;  T := T - {u};  w := [u,z];
        if w ∉ N then
          sol_normal_closure( G, N, w );
          V := {};  M := N;
          for each v in U do
            if v ∉ M then
              M := <M,v> using Algorithm 2; V := V ∪ {v};
            else
              T := T ∪ {v};
            end if;
          end for;
          U := V;
          if z ∈ M then
            done := true; break out of while T not empty loop;
          end if;
        end if; (* w ∉ N *)
      end while; (* T not empty *)
      if not done then
        M := <M,z> using Algorithm 2; U := U ∪ {z};
        Z := Z ∪ { s^-1 × z × s : s is a generator of G };
      end if; (* not done *)
    end if; (* z ∉ M *)
  end while; (* Z not empty *)
  N := M;
end (* sol_normal_closure *);
```

For example, consider the group G = <(1,2),(1,2,3,4)>, the symmetric group of degree 4. The first call to *sol_normal_closure* computes the normal closure of <(1,2)>. Initially, N and M are both trivial, and y = (1,2), the first generator. The first iteration of the outer **while**-loop has z=(1,2) and U empty, so M is updated by Algorithm 2 to <(1,2)>, and the elements $z^{(1,2)}$ = *identity* and $z^{(1,2,3,4)}$ = (2,3) are added to Z. Let us ignore the identity element as it is in M, so the next iteration of the outer **while**-loop has z=(2,3) and U={(1,2)}. The commutator w = [(1,2), (2,3)] = (1,2,3), which is not in N, so we recursively compute N = normal closure of <(1,2,3)> in G. When we return from the recursive call, N is the alternating group of degree 4, and M is updated to G. Hence, $z \in M$ and the algorithm is finished.

The recursive call initially has N and M trivial, and $y = (1,2,3)$. The first iteration of the outer **while**-loop updates M to be the cyclic group $<(1,2,3)>$ and $U = \{(1,2,3)\}$, and $Z = \{(2,3,4)\}$ (ignoring (1,3,2), the conjugate of (1,2,3) by (1,2), because it is in M). The next iteration of the outer **while**-loop checks the commutator $w = [\ (1,2,3),\ (2,3,4)\ ] = (1,4)(2,3)$. The commutator is not in N, so *sol_normal_closure* is called recursively to form the normal closure of $<(1,4)(2,3)>$ in G. When the call returns, N is the Klein four-group $<(1,4)(2,3),(1,3)(2,4)>$, and M is updated to be the alternating group of degree 4. The element $z = (2,3,4)$ is in M, so the algorithm is finished.

The recursive call for y=(1,4)(2,3) first creates the cyclic group $<(1,4)(2,3)>$ and sets $U = \{(1,4)(2,3)\}$ and $Z = \{(1,3)(2,4),(1,2)(3,4)\}$. The commutator of u=(1,4)(2,3) and z=(1,3)(2,4) is trivial, so Algorithm 2 updates M to $<(1,4)(2,3),(1,3)(2,4)>$, the Klein four-group, $U = \{ (1,4)(2,3), (1,3)(2,4) \}$, and $Z = \{ (1,2)(3,4), (1,4)(2,3), (1,3)(2,4) \}$. All the elements of Z are in M, so the algorithm quickly terminates.

The depth of recursion of Algorithm 12 is bounded by the length of the derived series of G. Bounds (in terms of the degree $|\Omega|$) are known on the length of the derived series of a soluble permutation group of degree $|\Omega|$. This information can be used to generalize the algorithm to apply even when we are not certain that the group is soluble. If the group is soluble, then the algorithm will terminate normally. If the group is not soluble then the bound on the depth of recursion will be exceeded.

Conditioned PC Presentation of a p-Group

The subnormal series associated with a conditioned pcp of a p-group must be prime-step and central. We have seen how to achieve the first criterion. The second criterion requires that

$$j > i \text{ implies } [g_j, g_i] \in G(j+1).$$

So we ensure the criterion is satisfied by repeatedly testing if the commutator $[g_j, g_i] \in G(j+1)$ and whenever the commutator is not in the j+1-st term of the series, then the series is modified by placing $[g_j, g_i]$ as the new g_j (and eliminating some now redundant generator). This process, described in Algorithm 13, converges to a central series.

A group of order p^2 is abelian, so we always have $[g_j, g_{j-1}] \in G(j+1)$.

Algorithm 13 : Central Series of a p-group

```
Input: a permutation p-group G with a base B;
    a prime-step B-ssgs [g_1, g_2, ..., g_m] of G;
Output: a prime-step B-ssgs [g_1, g_2, ..., g_m] of G with a central series;
begin
  for j := m downto 3 do
    for i := j-2 downto 1 do
      while not [g_j,g_i] ∈ <g_{j+1}, g_{j+2}, ..., g_m> do
        g := [ g_j, g_i ]; k := li( g ); (* the leading index of g as a normal word *);
        delete g_k;      insert g between g_{j+1} and g_j;
      end while;
    end for;
  end for;
end.
```

To preserve the B-strong property, the insertion of the commutator is done by repeatedly transposing elements (beginning at level k) until g is in the correct position. The B-ssgs change algorithm is used for each transposition.

As an example of Algorithm 13 consider the symmetries G of the square which has a base $B=[1,3]$ and a B-ssgs $\vec{g}$ given by $g_1=(3,4)$, $g_2=(1,2)(3,4)$, and $g_3=(1,3)(2,4)$. The commutator $[g_3,g_1]$ is g_2, so we change the B-ssgs to $[g_1,g_3,g_2]$. Now the commutator is the identity, and the subnormal series is central.

Conditioned PC Presentation of a Soluble Group

A conditioned pc presentation of a soluble group requires that the series associated with the pcp refines a normal series with elementary abelian factors. The easiest way to achieve this is for the normal series to be a refinement of the derived series. The factors of the derived series are abelian, and the abelian factors can be refined to elementary abelian ones (as in Algorithm 14).

Algorithm 14 : Refining an Abelian Factor

```
Input: a group G with a normal subgroup N such that G/N is abelian;
    a prime-step B-ssgs g⃗ = [g_1, g_2, ..., g_m] which extends N to G;

Output: a refinement of G/N into elementary abelian factors
        and the corresponding B-ssgs;
begin
  while N ≠ G do
    choose a prime p dividing the order of G/N;
    M := N;
    for each g in g⃗ but not in N do
      if g^p ∈ N do
        M := <M,g> using Algorithm 1;
      end if;
    end for;
    N := M; (* the next term in the series with e.a. factors *)
  end while;
end.
```

The soluble Schreier method can be modified to keep track of the derived series and return a B-ssgs with an associated subnormal series that refines the derived series. The modifications are presented in Algorithm 15.

Algorithm 15 : Modified Soluble Schreier Method

Input: a soluble permutation group G given by generators S;
Output: a base B for G and a B-ssgs $\vec{g}$ of G such that
the series refines the derived series of G;

```
begin
  N := < identity >;   B := empty;   g⃗ := empty;
  for each s in S do
    derived_normal_closure( G, N, s );
  end for;
  (* N is G *)
end.

procedure derived_normal_closure( G : group; var N : group; y : element of G );

(* Compute M, the normal closure of <N,y> in G where N ◁ G.
   We always have M/N is abelian during the course of the procedure.
   Assume we know the derived series of N and a corresponding base
   and B-ssgs. We maintain D(M) = D(N).
   At the end of the procedure, M is returned as N. *)

begin
  M := N;   U := generators extending D(N) to N;   Z := {y};
  while Z is not empty do
    choose z ∈ Z;   Z := Z - {z}; T := U;
    while T is not empty do
      choose u ∈ T;   T := T - {u};   w := [u,z];
      if w ∉ D(N) then
        derived_normal_closure( G, D(N), w ); M := D(N);
        V := {};
        for each v in U do
          if v ∉ M then
            M := <M,v> using Algorithm 2; V := V ∪ {v};
          end if;
        end for;
        U := V; (* generators extending D(N) = D(M) to M *)
      end if; (* w ∉ N *)
    end while; (* T not empty *)
    if not z ∈ M then
      M := <M,z> using Algorithm 2; U := U ∪ {z};
      Z := Z ∪ { s^-1 × z × s : s is a generator of G };
    end if;
  end while; (* Z not empty *)
  N := M;
end (* derived_normal_closure *);
```

Summary

This chapter has introduced the data structures of B-strong series-generating sequence and cyclically extended Schreier vectors as a representation of a soluble permutation group. Algorithms to construct the data structure, compute normal words, and compute with the isomorphism between the permutation group and the group defined by the pc presentation have been presented.

Exercises

(1/Easy) Consider the group G of order 2^7 and degree 8 generated by

g_7=(2,6)(4,5),
g_6=(4,5)(7,8),
g_5=(1,3)(2,6)(4,5)(7,8),
g_4=(4,7)(5,8),
g_3=(1,2)(3,6)(4,7)(5,8),
g_2=(1,7)(2,4)(3,8)(5,6),
g_1=(7,8).

The group has a base B=[1,2,4,7] and the sequence $[g_1,g_2,\ldots,g_7]$ is a B-ssgs.

(a) Construct the cyclically extended Schreier vectors of G.

(b) Form the normal word for y=(1,4,6,8,3,5,2,7).

(c) Let y=(1,7,6,5,3,8,2,4), and let $H = G(2) = \langle g_2,g_3,\ldots,g_7\rangle$. The square of y lies in H. Use Algorithm 1 to construct a B-ssgs of $\langle H,y\rangle$.

(d) Let y=(1,7,6,5,3,8,2,4), and let $H = G(5) = \langle g_5,g_6,g_7\rangle$. Use Algorithm 2 to construct a B-ssgs of $\langle H,y\rangle$.

(2/Moderate) For the group G of Exercise 1 with the given base B and B-ssgs, perform a sequence of changes of B-ssgs (by transposing adjacent members of the B-ssgs) until the final B-ssgs is $[g_1,g_4,g_6,g_7,g_2,g_3,g_5]$.

(3/Easy) Let $G = \langle\, a=(1,2,3,4)(5,6,7,8),\ b=(1,5,3,7)(2,8,4,6) \,\rangle$ of degree 8 and order 8. Execute Algorithm 7 to construct a base B and B-ssgs of G. Is the associated pc presentation conditioned?

(4. Moderate) For the group G of Exercise 1, perform Algorithm 7 and construct a base B and a B-ssgs of G.

Then perform Algorithm 13 (with appropriate uses of Algorithm 5) to construct a B-ssgs that gives a conditioned pc presentation of G (regarded as a p-group).

(5/Moderate) Let G be the soluble group of degree 6 and order 72 generated by

a=(1,2,3),
b=(2,3),
c=(1,4)(2,6,3,5).

(a) Execute Algorithm 12 to construct a base B=[1,2,4,5] and B-ssgs of G.

(b) Execute Algorithm 15 and Algorithm 14 to construct a B-ssgs of G such that the associated pc presentation is conditioned.

(c) Execute Algorithm 9 with G as the input group.

(d) Execute Algorithm 10 to find a normal subgroup M of G of prime index.

(6/Difficult) A complete labelled branching combines a set of Schreier vectors for a stabilizer chain into a single data structure, while a cesv combines the Schreier vectors of a subnormal series for a fixed level of their stabilizer chains. Can the two data structures be reconciled into one for a soluble permutation group?

Bibliographical Remarks

The first success in converting between a permutation representation and a pc presentation was by D.F. Holt in 1982 as part of his work on Schur multipliers: D.F. Holt, *"A computer program for the calculation of the Schur multiplier of a permutation group"*, **Computational Group Theory,** M.D. Atkinson (ed.), Academic Press, Academic Press, 1984, pp. 307-319. Given a B-ssgs of a Sylow p-subgroup, his algorithm produced a conditioned pc presentation by making the subnormal series prime-step and central. He also realised that cyclically extended Schreier vectors could represent all groups in the series and produce a normal word for a permutation. A proof of the algorithm is given in G. Butler, "*A proof of Holt's algorithm*", J. Symbolic Comp. **5** (1988) 275-283. The final step for permutation p-groups of constructing a B-ssgs from an arbitrary base and strong generating set was done by G. Butler and J.J. Cannon, "*On Holt's algorithm*", J. Symbolic Comp. to appear.

The search for generalizations of these techniques to soluble groups began in 1985 with G. Butler, "*Data structures and algorithms for cyclically extended Schreier vectors*", Congressus Numerantum **52** (1986) 63-78, which started with a derived series. The derived series was (is) too expensive to compute. That cost altered with the soluble Schreier method of Sims in 1987 published as C.C. Sims, "*Computing the order of a solvable permutation group*", J. Symbolic Comp. **9** (1990) 699-705. Sims' algorithm computes a B-ssgs, and can be modified so that the corresponding pc presentation of the soluble group is conditioned. The modifications were done by G. Butler, "*Computing a conditioned pc presentation of a soluble permutation group*", TR 392, Basser Department of Computer Science, University of Sydney, 1990, 5 pages.

Also in 1985, W.M. Kantor, "*Sylow's theorem in polynomial time*", J. Comput. System Sci. **30** (1985) 359-394, in the appendix, presents an algorithm for finding a normal subgroup of prime index in a soluble permutation group. The algorithm uses homomorphisms to reduce to the primitive case where the structure of the group is known. His treatment of the primitive case completes Algorithm 11 for constructing a B-ssgs in a way which is analogous to Algorithm 7 for permutation p-groups. The implementation details (mainly to avoid calls to Schreier-Sims method) are presented in G. Butler, "*Implementing some algorithms of Kantor*", 1991.

The structure of a primitive soluble group is discussed in Satz 3.2 of B. Huppert, **Endliche Gruppen I**, Springer-Verlag, Berlin, 1967, p.159. Bounds on the order of such groups are given by L. Palfy, "*A polynomial bound on the order of primitive soluble groups*", J. Algebra **77** (1982) 127-137 and T.R. Wolf, "*Solvable and nilpotent subgroups of $GL(n,q^m)$*", Canadian

J. Math. **34** (1982) 1097-1111. J.D. Dixon, "*The solvable length of a solvable linear group*", Math. Zeitschrift **107** (1968) 151-158, bounds the derived length of a soluble permutation group of degree $|\Omega|$ by $2.5 \log_3(|\Omega|)$.

The computation of $O_p(G)$ is presented in G. Butler and J.J. Cannon, "*Computing with permutation and matrix groups III: Sylow subgroups*", J. Symbolic Comp. **8** (1989) 241-252, while P.M. Neumann, "*Some algorithms for computing with permutation groups*", **Groups - St Andrews 1985**, E.F. Robertson and C.M. Campbell (eds), London Mathematics Society Lecture Notes **121**, Cambridge University Press, Cambridge, 1986, pp. 59-92, discusses an alternative approach to constructing an elementary abelian regular normal subgroup of a primitive group.

The series-generating sequences are called "smooth generating sequences" by Z. Galil, C.M. Hoffman, E.M. Luks, C.P. Schnorr, and A. Weber, "*An* $O(n^3 \log n)$ *deterministic and an* $O(n^3)$ *Las Vegas isomorphism test for trivalent graphs*", Journal of the ACM **34**, 3 (1987) 513-531, who use them to describe 2-groups given by permutations.

Chapter 19. Some Other Algorithms

This chapter presents a brief discussion of algorithms for permutation groups which are not treated in the book. The algorithms and the ideas behind them are important, but not fundamental enough to warrant a detailed presentation in this book. Pointers to the literature are given.

Single Coset Enumeration

There are several different kinds of problems addressed when enumerating the (left or right) cosets of a subgroup H of a permutation group G.

Problem 1: Given a permutation $g \in G$, determine a canonical representative of the coset $g \times H$.

Problem 2: Determine a set of coset representatives of H in G.

Problem 3: Determine the action of G on the cosets of H in G. That is, provide a permutation representation of G of degree $|G:H|$ such that H is isomorphic to the point stabiliser in this representation.

A solution to Problem 1 provides a solution to the other problems. However, one can often solve Problem 3 without defining a canonical coset representative.

The first approach for permutation groups relies on the ordering of elements of G induced from an ordering of the points such that the base points come first in the ordering. The canonical coset representative is the first element in the coset, and is formed by a process similar to testing membership. It is discussed in C.C. Sims, "*Computation with permutation groups*", **SYMSAM '71** (Proc. 2nd Symp. Symbolic and Algebraic Manipulation, Los Angeles, 1971), S.R. Petrick (ed.), ACM, New York, 1971, pp.23-28; J.S. Richardson, **GROUP: A computer system for group-theoretical calculations.** M.Sc. Thesis, University of Sydney, 1973; G. Butler, "*Effective computation with group homomorphisms*", J. Symbolic Comp., **1** (1985) 143-157.

Other orderings of the elements allow one to inductively work up the stabiliser chain, see D.F. Holt, "*The calculation of the Schur multiplier of a permutation group*", **Computational Group Theory,** M.D. Atkinson (ed.), Academic Press, Academic Press, 1984, pp.307-319, for hints.

A special case, important in enumerating combinatorial objects, has G equal to the symmetric group-see M. Furst, J.E. Hopcroft, and E. Luks, "*Polynomial-time algorithms for permutation groups*", Proc. 21st IEEE Foundations of Computer Science, 1980, pp.36-41; L. Allison, "*Generating coset representatives for permutation groups*", J. Algorithms **2** (1981) 227-244; M. Jerrum, "*A compact representation for permutation groups*", Proc. 23rd IEEE Foundations of Comp. Science, 1982, pp.126-133. (and J. Algorithms **7** (1986) 60-78.) Jerrum formulates Problem 1 and 2 in terms of topological sorts of a complete labelled branching of H.

Recently, the problems have been addressed by working up chains of subgroups involving set or block stabilisers: J.D. Dixon, and A. Majeed, "*Coset representatives for permutation groups*", Portugaliae Mathematica **45**, 1 (1988) 61-68; Wang DaFang, "*A new method for computation of permutation representations of a finite group*", manuscript, 1988.

A very space-efficient solution to Problem 3 is described in G. Cooperman and L. Finkelstein, "*New methods for using Cayley graphs in interconnection networks*", to appear in Discrete Applied Mathematics.

Double Coset Enumeration

There are several different kinds of problems addressed when enumerating the double cosets $H \times g \times K$ of subgroups H and K of a permutation group G.

Problem 1: Given a permutation $g \in G$, determine a canonical representative of the coset $H \times g \times K$.

Problem 2: Determine a set of double coset representatives of H and K in G.

The obvious way to approach the task is to solve Problem 3 of the single cosets of H, and then form the orbits of K. Each orbit of K in this representation corresponds to a double coset. This approach is used in D.F. Holt, "*The calculation of the Schur multiplier of a permutation group*", **Computational Group Theory,** M.D. Atkinson (ed.), Academic Press, Academic Press, 1984, pp.307-319.

Problem 1 has been addressed in G. Butler, "*On computing double coset representatives in permutation groups*", **Computational Group Theory,** M.D. Atkinson (ed.), Academic Press, Academic Press, 1984, pp. 283-290, where the canonical representative is the first element in the coset under the usual induced lexicographical ordering.

Several papers deal with the special case where G is the symmetric group, and the subgroups H and/or K are Young subgroups (that is, partition stabilisers in the symmetric group). H. Brown, L. Hjelmeland, and L. Masinter, "*Constructive graph labeling using double cosets*", Discrete Mathematics **7** (1974) 1-30; H. Brown, "*Molecular structure elucidation III*", SIAM J. Applied Math. **32**, 3 (1977) 534-551; R. Grund, **Computerunterstützte Konstruktion von speziellen Doppelnebenklassentransversalen und deren Anwendungen auf die konstruktive Kombinatorik**, Diplomarbeit, Universität Bayreuth, 1989; R. Grund, "*Symmetrieklassen von Abbildungen und die Konstruktion von diskreten Strukturen*", Bayreuther Mathematische Schriften **31** (1990) 19-54. B. Schmalz, **Computerunterstützte Konstruktion von Doppelnebenklassenrepräsentanten mit Anwendungen auf das Isomorphieproblem der Graphentheorie**, Diplomarbeit, Universität Bayreuth, 1989; B. Schmalz, "*Verwendung von Untergruppenleitern zur Bestimmung von Doppelnebenklassen*", Bayreuther Mathematische Schriften **31** (1990) 109-143.

Verify

An algorithm to verify that a base and strong generating set is correct was developed by Sims during the construction of several sporadic simple groups. It has been implemented by John Brownie and John Cannon to handle more general cases, and is effective for permutation groups of degree one million. See C.C. Sims, "*A method for constructing a group from a subgroup*", **Topics in Algebra,** (Proceedings of 18th Summer Research Institute, Canberra, 1978), M.F. Newman (ed.), Lecture Notes in Mathematics **697**, Springer-Verlag, Berlin, 1978, and forthcoming papers.

Conjugacy Classes of Elements

The conjugacy classes of elements of a group *G* plays a critical role in many algorithms concerned with the global structure of a group, such as algorithms for computing the lattice of subgroups or normal subgroups, the character table, the automorphism group, and the maximal subgroups.

For small groups, the obvious approach is to consider the group *G* acting on its elements by conjugation, and to compute the orbits.

For highly transitive groups, a method of Sims computes representatives of the orbits of *G* on *k*-tuples (for small values of *k*), and then considerings brackettings of the points in the tuple as cycles of possible canonical representatives of the conjugacy classes. See C.C. Sims, "*Determining the conjugacy classes of a permutation group*", **Computers in Algebra and Number Theory,** G. Birkhoff and M. Hall, Jr (eds), SIAM-AMS Proc., **4**, Amer. Math. Soc., Providence, R.I., 1971, pp.191-195. The method was implemented and studied by G. Butler, **Computational Approaches to Certain Problems in the Theory of Finite Groups,** Ph. D. Thesis, University of Sydney, 1980, but does not appear to be generally effective.

The thesis also introduces a random method for finding class representatives that is effective for groups of moderate degree and order which are close to being simple.

An inductive approach, which uses a lot of group-theoretic knowledge and algorithms is described in G. Butler, "*An inductive schema for computing conjugacy classes in permutation groups*", TR 394, Basser Department of Computer Science, University of Sydney, 1990.

Cohomology

The work of Derek Holt on cohomology has integrated many machine representations of groups to tackle a complex problem, and has led to several new algorithms that have fundamental significance : D.F. Holt, "*A computer program for the calculation of the Schur multiplier of a permutation group*", **Computational Group Theory,** M.D. Atkinson (ed.), Academic Press, Academic Press, 1984, pp. 307-319; D.F. Holt, "*A computer program for the calculation of a covering group of a finite group*", J. Pure Applied Algebra **35** (1985) 287-295; D.F. Holt, "*The mechanical computation of first and second cohomology groups*", J. Symbolic Comp. **1** (1985) 351-361.

Group Recognition

Given a set of generators for a group *G*, it is useful to be able to recognise whether *G* is the alternating or symmetric group of the same degree, because these groups are very large and may be the worst cases for the algorithm one is about to apply (even though the answer in these cases may be obvious). Much work has been done on this problem, see J.J. Cannon, "*A computational toolkit for finite permutation groups*", **Proceedings of the Rutgers Group Theory Year, 1983-1984,** M. Aschbacher, D. Gorenstein, R. Lyons, M. O'Nan, C. Sims, W. Feit (editors), CUP, New York, 1984, pp.1-18.

It is also useful to be able to recognize a group *G* as an explicit abstract group in some classification, such as the doubly-transitive groups, or primitive groups, or simple groups. In the latter case, one may first have to show the group is simple. These problems are addressed in J.J. Cannon, "*Effective procedures for the recognition of primitive groups*", **The Santa Cruz Conference on Finite Groups,** AMS Proc. Symp. Pure Mathematics **37** (1980) 487-493; P.M. Neumann, *"Some algorithms for computing with permutation groups"*, **Groups - St Andrews 1985**, E.F. Robertson and C.M. Campbell (eds), London Mathematics Society Lecture Notes **121**, Cambridge University Press, Cambridge, 1986, pp. 59-92.

Composition factors

The composition factors of a group *G* are the simple groups which occur as factors groups *H/N*, for some subgroup *H* of *G* and normal subgroup *N* of *H*. The algorithms reduce to the primitive case, and then utilise the O'Nan-Scott Theorem and the classification of finite simple groups. Of theoretical interest is E.M. Luks, "*Computing the composition factors of a permutation group in polynomial time*", Combinatorica **7** (1987) 87-99. Implementations by John Cannon follow the algorithms described in P.M. Neumann, *"Some algorithms for computing with permutation groups"*, **Groups - St Andrews 1985**, E.F. Robertson and C.M. Campbell (eds), London Mathematics Society Lecture Notes **121**, Cambridge University Press, Cambridge, 1986, pp. 59-92; and more recently, the algorithm of W.M. Kantor, "*Finding composition factors of permutation groups of degree* 10^6", to appear in J. Symbolic Comp.

Some Ideas From Complexity Results

There are several exciting advances in the complexity of group theoretic problems, that may indeed impact upon methods for effective computation. We highlight those we consider most significant or most accessible to the reader: L. Babai, E. Luks, A. Seress, "*Fast management of permutation groups*", Proc. 28th IEEE Symposium on Foundations of Computer Science, 1988, pp.272-282; G. Cooperman, L. Finkelstein and E. M. Luks, "*Reduction of group constructions to point stabilizers*", **ISSAC 89**, (Proc. 1989 ACM Symposium on Symbolic and Algebraic Computation, Portland, Oregon), ACM Press, New York, 1989, pp.351-356; W.M. Kantor and E.M. Luks, "*Computing in quotient groups*", Proc. 22nd ACM Symposium on Theory of Computing, Baltimore, 1990, pp.524-534; L. Babai, "*Local expansion of vertex-transitive graphs and random generation in finite groups*", Proc. 23nd ACM Symposium on Theory of Computing, 1991, to appear; L. Babai, G. Cooperman, L. Finkelstein, E. Luks, A. Seress, "*Fast Monte Carlo algorithms for permutation groups*", Proc. 23nd ACM Symposium on Theory of Computing, 1991, to appear.

Index of Algorithms

10. Inductive Foundation

11. Backtrack Search

12. Base Change

13. Schreier-Sims Method

14. Complexity of the Schreier-Sims Method

15. Homomorphisms

16. Sylow Subgroups

17. P-Groups and Soluble Groups

18. Soluble Permutation Groups

Index of Definitions

Lecture Notes in Computer Science

For information about Vols. 1–473
please contact your bookseller or Springer-Verlag

Vol. 474: D. Karagiannis (Ed.), Information Systems and Artificial Intelligence: Integration Aspects. Proceedings, 1990. X, 293 pages. 1991. (Subseries LNAI).

Vol. 475: P. Schroeder-Heister (Ed.), Extensions of Logic Programming. Proceedings, 1989. VIII, 364 pages. 1991. (Subseries LNAI).

Vol. 476: M. Filgueiras, L. Damas, N. Moreira, A.P. Tomás (Eds.), Natural Language Processing. Proceedings, 1990. VII, 253 pages. 1991. (Subseries LNAI).

Vol. 477: D. Hammer (Ed.), Compiler Compilers. Proceedings, 1990. VI, 227 pages. 1991.

Vol. 478: J. van Eijck (Ed.), Logics in AI. Proceedings, 1990. IX, 562 pages. 1991. (Subseries in LNAI).

Vol. 479: H. Schmidt, Meta-Level Control for Deductive Database Systems. VI, 155 pages. 1991.

Vol. 480: C. Choffrut, M. Jantzen (Eds.), STACS 91. Proceedings, 1991. X, 549 pages. 1991.

Vol. 481: E. Lang, K.-U. Carstensen, G. Simmons, Modelling Spatial Knowledge on a Linguistic Basis. IX, 138 pages. 1991. (Subseries LNAI).

Vol. 482: Y. Kodratoff (Ed.), Machine Learning – EWSL-91. Proceedings, 1991. XI, 537 pages. 1991. (Subseries LNAI).

Vol. 483: G. Rozenberg (Ed.), Advances in Petri Nets 1990. VI, 515 pages. 1991.

Vol. 484: R. H. Möhring (Ed.), Graph-Theoretic Concepts in Computer Science. Proceedings, 1990. IX, 360 pages. 1991.

Vol. 485: K. Furukawa, H. Tanaka, T. Fuijsaki (Eds.), Logic Programming '89. Proceedings, 1989. IX, 183 pages. 1991. (Subseries LNAI).

Vol. 486: J. van Leeuwen, N. Santoro (Eds.), Distributed Algorithms. Proceedings, 1990. VI, 433 pages. 1991.

Vol. 487: A. Bode (Ed.), Distributed Memory Computing. Proceedings, 1991. XI, 506 pages. 1991.

Vol. 488: R. V. Book (Ed.), Rewriting Techniques and Applications. Proceedings, 1991. VII, 458 pages. 1991.

Vol. 489: J. W. de Bakker, W. P. de Roever, G. Rozenberg (Eds.), Foundations of Object-Oriented Languages. Proceedings, 1990. VIII, 442 pages. 1991.

Vol. 490: J. A. Bergstra, L. M. G. Feijs (Eds.), Algebraic Methods II: Theory, Tools and Applications. VI, 434 pages. 1991.

Vol. 491: A. Yonezawa, T. Ito (Eds.), Concurrency: Theory, Language, and Architecture. Proceedings,1989. VIII, 339 pages. 1991.

Vol. 492: D. Sriram, R. Logcher, S. Fukuda (Eds.), Computer-Aided Cooperative Product Development. Proceedings, 1989 VII, 630 pages. 1991.

Vol. 493: S. Abramsky, T. S. E. Maibaum (Eds.), TAPSOFT '91. Volume 1. Proceedings, 1991. VIII, 455 pages. 1991.

Vol. 494: S. Abramsky, T. S. E. Maibaum (Eds.), TAPSOFT '91. Volume 2. Proceedings, 1991. VIII, 482 pages. 1991.

Vol. 495: 9. Thalheim, J. Demetrovics, H.-D. Gerhardt (Eds.), MFDBS '91. Proceedings, 1991. VI, 395 pages. 1991.

Vol. 496: H.-P. Schwefel, R. Männer (Eds.), Parallel Problem Solving from Nature. Proceedings, 1990. XI, 485 pages. 1991.

Vol. 497: F. Dehne, F. Fiala. W.W. Koczkodaj (Eds.), Advances in Computing and Information - ICCI '91. Proceedings, 1991. VIII, 745 pages. 1991.

Vol. 498: R. Andersen, J. A. Bubenko jr., A. Sølvberg (Eds.), Advanced Information Systems Engineering. Proceedings, 1991. VI, 579 pages. 1991.

Vol. 499: D. Christodoulakis (Ed.), Ada: The Choice for '92. Proceedings, 1991. VI, 411 pages. 1991.

Vol. 500: M. Held, On the Computational Geometry of Pocket Machining. XII, 179 pages. 1991.

Vol. 501: M. Bidoit, H.-J. Kreowski, P. Lescanne, F. Orejas, D. Sannella (Eds.), Algebraic System Specification and Development. VIII, 98 pages. 1991.

Vol. 502: J. Bārzdiņž, D. Bjørner (Eds.), Baltic Computer Science. X, 619 pages. 1991.

Vol. 503: P. America (Ed.), Parallel Database Systems. Proceedings, 1990. VIII, 433 pages. 1991.

Vol. 504: J. W. Schmidt, A. A. Stogny (Eds.), Next Generation Information System Technology. Proceedings, 1990. IX, 450 pages. 1991.

Vol. 505: E. H. L. Aarts, J. van Leeuwen, M. Rem (Eds.), PARLE '91. Parallel Architectures and Languages Europe, Volume I. Proceedings, 1991. XV, 423 pages. 1991.

Vol. 506: E. H. L. Aarts, J. van Leeuwen, M. Rem (Eds.), PARLE '91. Parallel Architectures and Languages Europe, Volume II. Proceedings, 1991. XV, 489 pages. 1991.

Vol. 507: N. A. Sherwani, E. de Doncker, J. A. Kapenga (Eds.), Computing in the 90's. Proceedings, 1989. XIII, 441 pages. 1991.

Vol. 508: S. Sakata (Ed.), Applied Algebra, Algebraic Algorithms and Error-Correcting Codes. Proceedings, 1990. IX, 390 pages. 1991.

Vol. 509: A. Endres, H. Weber (Eds.), Software Development Environments and CASE Technology. Proceedings, 1991. VIII, 286 pages. 1991.

Vol. 510: J. Leach Albert, B. Monien, M. Rodríguez (Eds.), Automata, Languages and Programming. Proceedings, 1991. XII, 763 pages. 1991.

Vol. 511: A. C. F. Colchester, D.J. Hawkes (Eds.), Information Processing in Medical Imaging. Proceedings, 1991. XI, 512 pages. 1991.

Vol. 512: P. America (Ed.), ECOOP '91. European Conference on Object-Oriented Programming. Proceedings, 1991. X, 396 pages. 1991.

Vol. 513: N. M. Mattos, An Approach to Knowledge Base Management. IX, 247 pages. 1991. (Subseries LNAI).

Vol. 514: G. Cohen, P. Charpin (Eds.), EUROCODE '90. Proceedings, 1990. XI, 392 pages. 1991.

Vol. 515: J. P. Martins, M. Reinfrank (Eds.), Truth Maintenance Systems. Proceedings, 1990. VII, 177 pages. 1991. (Subseries LNAI).

Vol. 516: S. Kaplan, M. Okada (Eds.), Conditional and Typed Rewriting Systems. Proceedings, 1990. IX, 461 pages. 1991.

Vol. 517: K. Nökel, Temporally Distributed Symptoms in Technical Diagnosis. IX, 164 pages. 1991. (Subseries LNAI).

Vol. 518: J. G. Williams, Instantiation Theory. VIII, 133 pages. 1991. (Subseries LNAI).

Vol. 519: F. Dehne, J.-R. Sack, N. Santoro (Eds.), Algorithms and Data Structures. Proceedings, 1991. X, 496 pages. 1991.

Vol. 520: A. Tarlecki (Ed.), Mathematical Foundations of Computer Science 1991. Proceedings, 1991. XI, 435 pages. 1991.

Vol. 521: B. Bouchon-Meunier, R. R. Yager, L. A. Zadek (Eds.), Uncertainty in Knowledge-Bases. Proceedings, 1990. X, 609 pages. 1991.

Vol. 522: J. Hertzberg (Ed.), European Workshop on Planning. Proceedings, 1991. VII, 121 pages. 1991. (Subseries LNAI).

Vol. 523: J. Hughes (Ed.), Functional Programming Languages and Computer Architecture. Proceedings, 1991. VIII, 666 pages. 1991.

Vol. 524: G. Rozenberg (Ed.), Advances in Petri Nets 1991. VIII, 572 pages. 1991.

Vol. 525: O. Günther, H.-J. Schek (Eds.), Advances in Spatial Databases. Proceedings, 1991. XI, 471 pages. 1991.

Vol. 526: T. Ito, A. R. Meyer (Eds.), Theoretical Aspects of Computer Software. Proceedings, 1991. X, 772 pages. 1991.

Vol. 527: J.C.M. Baeten, J. F. Groote (Eds.), CONCUR '91. Proceedings, 1991. VIII, 541 pages. 1991.

Vol. 528: J. Maluszynski, M. Wirsing (Eds.), Programming Language Implementation and Logic Programming. Proceedings, 1991. XI, 433 pages. 1991.

Vol. 529: L. Budach (Ed.), Fundamentals of Computation Theory. Proceedings, 1991. XII, 426 pages. 1991.

Vol. 530: D. H. Pitt, P.-L. Curien, S. Abramsky, A. M. Pitts, A. Poigné, D. E. Rydeheard (Eds.), Category Theory and Computer Science. Proceedings, 1991. VII, 301 pages. 1991.

Vol. 531: E. M. Clarke, R. P. Kurshan (Eds.), Computer-Aided Verification. Proceedings, 1990. XIII, 372 pages. 1991.

Vol. 532: H. Ehrig, H.-J. Kreowski, G. Rozenberg (Eds.), Graph Grammars and Their Application to Computer Science. Proceedings, 1990. X, 703 pages. 1991.

Vol. 533: E. Börger, H. Kleine Büning, M. M. Richter, W. Schönfeld (Eds.), Computer Science Logic. Proceedings, 1990. VIII, 399 pages. 1991.

Vol. 534: H. Ehrig, K. P. Jantke, F. Orejas, H. Reichel (Eds.), Recent Trends in Data Type Specification. Proceedings, 1990. VIII, 379 pages. 1991.

Vol. 535: P. Jorrand, J. Kelemen (Eds.), Fundamentals of Artificial Intelligence Research. Proceedings, 1991. VIII, 255 pages. 1991. (Subseries LNAI).

Vol. 536: J. E. Tomayko, Software Engineering Education. Proceedings, 1991. VIII, 296 pages. 1991.

Vol. 537: A. J. Menezes, S. A. Vanstone (Eds.), Advances in Cryptology – CRYPTO '90. Proceedings. XIII, 644 pages. 1991.

Vol. 538: M. Kojima, N. Megiddo, T. Noma, A. Yoshise, A Unified Approach to Interior Point Algorithms for Linear Complementarity Problems. VIII, 108 pages. 1991.

Vol. 539: H. F. Mattson, T. Mora, T. R. N. Rao (Eds.), Applied Algebra, Algebraic Algorithms and Error-Correcting Codes. Proceedings, 1991. XI, 489 pages. 1991.

Vol. 540: A. Prieto (Ed.), Artificial Neural Networks. Proceedings, 1991. XIII, 476 pages. 1991.

Vol. 541: P. Barahona, L. Moniz Pereira, A. Porto (Eds.), EPIA '91. Proceedings, 1991. VIII, 292 pages. 1991. (Subseries LNAI).

Vol. 543: J. Dix, K. P. Jantke, P. H. Schmitt (Eds.), Nonmonotonic and Inductive Logic. Proceedings, 1990. X, 243 pages. 1991. (Subseries LNAI).

Vol. 544: M. Broy, M. Wirsing (Eds.), Methods of Programming. XII, 268 pages. 1991.

Vol. 545: H. Alblas, B. Melichar (Eds.), Attribute Grammars, Applications and Systems. Proceedings, 1991. IX, 513 pages. 1991.

Vol. 547: D. W. Davies (Ed.), Advances in Cryptology – EUROCRYPT '91. Proceedings, 1991. XII, 556 pages. 1991.

Vol. 548: R. Kruse, P. Siegel (Eds.), Symbolic and Quantitative Approaches to Uncertainty. Proceedings, 1991. XI, 362 pages. 1991.

Vol. 550: A. van Lamsweerde, A. Fugetta (Eds.), ESEC '91. Proceedings, 1991. XII, 515 pages. 1991.

Vol. 551:S. Prehn, W. J. Toetenel (Eds.), VDM '91. Formal Software Development Methods. Volume 1. Proceedings, 1991. XIII, 699 pages. 1991.

Vol. 552: S. Prehn, W. J. Toetenel (Eds.), VDM '91. Formal Software Development Methods. Volume 2. Proceedings, 1991. XIV, 430 pages. 1991.

Vol. 553: H. Bieri, H. Noltemeier (Eds.), Computational Geometry - Methods, Algorithms and Applications '91. Proceedings, 1991. VIII, 320 pages. 1991.

Vol. 554: G. Grahne, The Problem of Incomplete Information in Relational Databases. VIII, 156 pages. 1991.

Vol. 555: H. Maurer (Ed.), New Results and New Trends in Computer Science. Proceedings, 1991. VIII, 403 pages. 1991.

Vol. 556: J.-M. Jacquet, Conclog: A Methodological Approach to Concurrent Logic Programming. XII, 781 pages. 1991.

Vol. 557: W. L. Hsu, R. C. T. Lee (Eds.), ISA '91 Algorithms. Proceedings, 1991. X, 396 pages. 1991.

Vol. 558: J. Hooman, Specification and Compositional Verification of Real-Time Systems. VIII, 235 pages. 1991.

Vol. 559: G. Butler, Fundamental Algorithms for Permutation Groups. XII, 238 pages. 1991.